Food Microbiology and Food Safety

Series Editor:
Michael P. Doyle

More information about this series at http://www.springer.com/series/7131

Food Microbiology and Food Safety Series

The Food Microbiology and Food Safety series is published in conjunction with the International Association for Food Protection, a non-profit association for food safety professionals. Dedicated to the life-long educational needs of its Members, IAFP provides an information network through its two scientific journals (Food Protection Trends and Journal of Food Protection), its educational Annual Meeting, international meetings and symposia, and interaction between food safety professionals.

Series Editor

Michael P. Doyle, *Regents Professor and Director of the Center for Food Safety, University of Georgia, Griffith, GA, USA*

Editorial Board

Judy A. Harrison

Editor

Food Safety for Farmers Markets: A Guide to Enhancing Safety of Local Foods

 Springer

Editor
Judy A. Harrison
Department of Foods and Nutrition
University of Georgia
Athens, GA, USA

Food Microbiology and Food Safety
ISBN 978-3-319-88302-1   ISBN 978-3-319-66689-1   (eBook)
DOI 10.1007/978-3-319-66689-1

Printed on acid-free paper

This Springer imprint is published by Springer Nature
The registered company is Springer International Publishing AG
The registered company address is: Gewerbestrasse 11, 6330 Cham, Switzerland

Preface

Interest in buying locally produced foods is a trend seen in many countries. In the USA, campaigns such as the US Department of Agriculture's (USDA) *Know Your Farmer, Know Your Food* and *The People's Garden* and the operation of the USDA Farmers' Market in Washington, DC, have helped to market the "local food" movement. In 1994, there were 1755 farmers' markets listed in the USDA Agricultural Marketing Service's (AMS) Farmers' Market Directory Listing. In 2016, there were 8669 markets listed. This represents an increase of almost 400% since the early 1990s. Similar trends can be seen in other countries as well with a 157% increase in markets in Australia between 2004 and 2015 and a range from 30% to 60% across several Canadian provinces in recent years.

In a 2011 report from the USDA Economic Research Service (ERS), small farms selling less than $50,000 in gross annual sales accounted for 81% of all farms reporting local food sales in 2008. The report stated that these farms were more likely to depend exclusively on direct-to-consumer sales at farmers' markets and roadside stands. In 2012, 70% of farms selling foods locally sold directly to consumers through farmers' markets and community-supported agriculture organizations (CSAs).

Although there have been few documented outbreaks or cases of illness attributed to farmers' markets, studies have identified a lack of food safety practices on small farms and in farmers' markets through both survey research and observational studies. The practices in use identified in these studies could increase consumers' risk for foodborne illnesses.

Studies from various countries examining why consumers shop at farmers' markets and roadside stands and through CSAs indicate that consumers want to meet and connect with the person who produced the food, to obtain higher-quality and fresher products, to have a healthier diet, to support the local community, to enjoy the social atmosphere of the farmers' market, to protect the environment, and to be safer from pesticides, added hormones, and foodborne illnesses.

The 2014 National Farmers' Market Manager Survey conducted by the USDA Agricultural Marketing Service (AMS) with 1400 farmers' market managers found that among 91% of those who managed markets in 2012 and 2013, over 60%

reported increases in the number of customers, increases in the number of repeat customers, and higher annual sales. Eighty-five percent of the managers surveyed were seeking to add vendors, with 62% of them seeking vendors selling different types of products, not just fresh fruits and vegetables.

Markets have increased interest in entrepreneurship and cottage food industries. Farmers' markets have become venues for products requiring various types of food safety training and licenses. Studies have found that requirements may vary from state to state or region to region and country to country. Some studies have identified a lack of awareness about regulations and a lack of knowledge about food safety principles that apply to foods being made and sold through very small businesses or under cottage food regulations, including knowledge about food allergens and labeling requirements.

Increasing numbers of customers buying locally in their quest for what they believe to be fresher, healthier, and safer foods create a challenge for food safety educators and for regulatory agencies. Farmers' markets offer unique shopping opportunities for consumers and meaningful opportunities for the farmer. It is important for public health and for the economic viability of farmers and farmers' markets to provide food using the best food safety practices. The purpose of this book is to provide an overview of potential food safety issues on farms and in markets and present best practices to enhance food safety at farms and farmers' markets both in the USA and internationally.

Athens, GA, USA   Judy A. Harrison

Contents

Contributors

Renee R. Boyer Department of Food Science and Technology, Virginia Tech, Blacksburg, VA, USA

Benjamin Chapman Department of Youth, Family and Community Sciences, North Carolina State University, Raleigh, NC, USA

Faith J. Critzer Department of Food Technology and Science, University of Tennessee, Knoxville, TN, USA

Jeff Farber Department of Food Science, University of Guelph, Guelph, ON, Canada

Judy A. Harrison Department of Foods and Nutrition, University of Georgia, Athens, GA, USA

Edward Jansson NSW Food Authority, Newington, NSW, Australia

Heather Lim Bureau of Food Surveillance and Science Integration, Health Canada/Government of Canada, Ottawa, ON, Canada

Bruce Nelan NSW Food Authority, Newington, NSW, Australia

Stephanie Pollard Department of Food Science and Technology, Virginia Tech, Blacksburg, VA, USA

Allison Sain Department of Food, Bioprocessing and Nutritional Sciences, North Carolina State University, Raleigh, NC, USA

Lisa Szabo NSW Food Authority, Newington, NSW, Australia

Lily L. Yang Department of Food Science and Technology, Virginia Tech, Blacksburg, VA, USA

Chapter 1
An Introduction to Microorganisms That Can Impact Products Sold at Farmers Markets

Faith J. Critzer

Abstract With the growing demand for fresh, locally sourced foods, the popularity of farmers markets have soared. Over the past decade, the number of farmers markets across the United States has steadily increased. With the growing popularity of these markets, increased awareness of food safety must also be considered. The food safety management practices used when growing and preparing foods for sale at the farmers market are based upon scientific principles. The goal of these practices are to decrease the likelihood food will be contaminated with harmful microorganisms that can make us ill, referred to as foodborne pathogens, or to inhibit bacterial foodborne pathogens from growing on the food prior to consumption. This chapter presents a broad overview of foodborne pathogens associated with products sold at farmers markets, sources of these pathogens, and means to inhibit bacterial growth.

Keywords Foodborne pathogens • pH • Water activity • Temperature • *Salmonella* • *Campylobacter*

The organisms that are likely to impact the safety of foods sold at farmers markets are similar to those we would expect for foods sold in any retail setting. Many individuals believe that since news reports of outbreaks linked to local farmers markets are relatively scarce to unheard of, foodborne pathogens are not an issue with foods grown and manufactured locally. However, pathogens do not preferentially select to contaminate food associated with large farms, food manufacturers, or retail operations.

Bellemare et al. have reported a positive correlation with the number of farmers markets per capita, the number of foodborne outbreaks, cases of campylobacteriosis (the illness caused by *Campylobacter* spp.) and outbreaks related to *Campylobacter* [1]. Based upon data from 2004 to 2011, a 1% increase in the number of farmers markets resulted in a 0.7% increase in foodborne outbreaks and 3.9% increase in

F.J. Critzer (✉)
Department of Food Technology and Science, University of Tennessee,
103 Food Science and Technology Building, 2600 River Drive, Knoxville, TN 37996, USA
e-mail: faithc@utk.edu

© Springer International Publishing AG 2017
J.A. Harrison (ed.), *Food Safety for Farmers Markets: A Guide to Enhancing Safety of Local Foods*, Food Microbiology and Food Safety,
DOI 10.1007/978-3-319-66689-1_1

foodborne illnesses. These increases were even more substantial when only evaluating *Campylobacter jejuni*, which showed a 3.9% increase for outbreaks under the same farmers market growth parameters. If a cause and effect relationship exists, it is important to understand what drivers may be at play resulting in the increase of food-borne illnesses. A possibility may be that food safety best practices are not uniformly practiced by farmers market vendors, which in many locations are self-governed or irregularly inspected for adherence. In reality, as presented in future chapters, many vendors do not have any mandatory regulations imposed upon them for safe growing, harvesting, or manufacturing of the foods they sell through these venues.

In order to reduce the risk of contamination, vendors must proactively adopt management practices that are based upon scientifically valid food safety principles. Limitations to epidemiological traceback to a causative agent associated with food-borne illnesses are a primary issue for local distribution systems since the popula-tion of people consuming these goods is relatively small, the probability of having enough people reporting illness will be even less and most likely result in what appears to be sporadic illnesses. While outbreaks linked to farmers markets are not as readily publicized, future chapters in this book will discuss some of the inci-dences of outbreaks linked to foods sold through this outlet.

Farmers market vendors must accept that without adherence to best practices, they are opening their operation up to the risk of making their patrons ill. This chap-ter presents an introduction to the types and properties of foodborne pathogens; and the burden of some foodborne pathogens, their characteristics, and sources where they are commonly found.

Types of Foodborne Pathogens

Foodborne pathogens are broadly categorized as disease-causing bacteria, viruses, or parasites associated with food. Each category has differences that should be well understood by those tasked with managing the safety of foods sold at farmers markets. Table 1.1 describes the overall burden from major bacterial, parasitic and viral foodborne pathogens with relation to illnesses and hospitalizations annually [2]. Collectively, this group is estimated to cause 9.4 million illnesses through con-tamination or intoxication of foods [2]. Estimates of the burden of foodborne patho-gens are used rather than actual reported cases because of drastic underreporting in the United States. This is due to numerous factors, including the self-limiting nature of many foodborne intoxications and infections because the ill person recovers rather quickly before they seek medical care [3]. Even if someone does seek medi-cal care, the physician must collect and submit a stool specimen, from which a foodborne pathogen must be isolated, followed by reporting from the physician to the public health officials. A breakdown at any one step will result in a lack of reporting, making it rather easy to understand why estimates give us a more reason-able understanding of the overall impact of foodborne pathogens.

Organisms which cause foodborne illness can be found in a number of places as shown in Table 1.2. The gastrointestinal tract is a primary source of many foodborne

Table 1.1 Estimates of foodborne illness caused by 31 major foodborne pathogens annually in the United States

Foodborne pathogen type	Ability to grow in foods	Estimated annual illnesses in the U.S. (% total) [2]	Estimated annual hospitalizations in the U.S. (% total) [2]	Examples
Bacteria	Yes	3.6 million (39%)	35,815 (64%)	*Campylobacter* spp.
				Listeria monocytogenes
				Salmonella
				Shiga toxigenic *Escherichia coli*
Protozoa	No	200,000 (2%)	5036 (9%)	*Giardia intestinalis*
				Cryptosporidium spp.
Viruses	No	5.5 million (59%)	15,109 (27%)	Hepatitis A
				Norovirus

pathogens. This is because most foodborne pathogens can be either symptomatically or asymptomatically carried in the gastrointestinal (G.I.) tract of humans and animals. From the G.I. tract they can be spread to many environmental sources, such as water and soil, where they can survive extremely long periods of time. This is one reason why fresh produce can easily become contaminated with foodborne pathogens via contact with the soil, animal manure, or contaminated water and is further described in Chaps. 2 and 4. People actively shedding foodborne pathogens either symptomatically or asymptomatically can also spread these pathogens to foods they contact. This has been observed as a route of contamination in retail settings as well as with vendors selling ready-to-eat foods at farmers markets. For these reasons, future chapters will cover segregation of ill workers both on the farm and at the market as a primary strategy to limit spread of foodborne pathogens.

Unlike viruses and parasites, bacteria have the ability to grow outside of a living host and can therefore multiply in number on foods or in the environment if conditions are favorable. Under the most favorable conditions, bacterial numbers can double in 20 min. The primary factors which affect the growth of bacterial foodborne pathogens include atmosphere, temperature, pH, water activity and availability of nutrients. Understanding these factors helps to explain strategies for controlling the growth of bacterial foodborne pathogens presented in future chapters. Viruses and parasites, including parasitic protozoa, must have a living host to replicate, such as when they are actively infecting humans.

Primary Factors Which Affect the Growth of Bacterial Foodborne Pathogens

Atmosphere

Bacteria are categorized by their ability to grow in the presence of oxygen. If an organism must have oxygen to grow, it is described as a strict aerobe. Given the fact that most bacteria that cause foodborne illness do so by infecting our

Table 1.2 Seven sources of pathogenic foodborne bacteria, protozoa, and viruses

	Soil and water	AIR AND DUST	Plants	Gastrointestinal tract	Food handlers	Animal feeds	Animal hides
Bacteria							
Bacillus	✓+[a]	✓+	✓		✓	✓	✓
Campylobacter jejuni and *coli*				✓+	✓		
Clostridium botulinum and *perfringens*	✓+[a]	✓+	✓	✓	✓	✓	✓
Escherichia coli	✓		✓	✓+	✓		
Listeria monocytogenes	✓		✓+		✓	✓	✓
Salmonella enterica				✓+		✓+	
Shigella sonnei and *flexneri*				✓+			
Staphylococcus aureus				✓	✓+		✓
Vibrio spp.	✓+[b]			✓			
Yersinia enterocolitica	✓		✓	✓			
Protozoa							
Cyclospora cayetanensis	✓		✓	✓			
Cryptosporidium parvum	✓+[b]			✓	✓		
Giardia lamblia	✓+[b]			✓	✓		
Toxoplasma gondii			✓	✓+			
Viruses							
Hepatitis A	✓+		✓	✓+	✓+		
Norovirus	✓		✓	✓+	✓+		

Note: + indicates a very important source
[a]Primarily soil
[b]Primarily water
Adapted from [4]

gastrointestinal tract, there are no foodborne pathogens which fall within this category. Rather, foodborne pathogens are categorized as strict anaerobes, meaning they cannot grow in the presence of oxygen or facultative anaerobes meaning they can grow without oxygen, but can increase their metabolic processes and grow faster if oxygen is present. *Clostridium* spp. are strict anaerobes and *Campylobacter* spp. are microaerophilic, meaning they need reduced oxygen content (3–5%) in order to grow. All other foodborne pathogens are considered to be facultative anaerobes. This is especially important when considering storage conditions of foods, such as those with vacuum packing or those which are filled hot and processed to remove oxygen in the headspace, such as with canned goods like sauces and pickles. Vacuum packaging and hot filling are two examples of how we have altered atmospheres of common foods in order to inhibit spoilage organisms. While this will help achieve a longer shelf-life, care must be taken to assure that foodborne pathogens cannot proliferate. This can be done through processing steps that will inactivate foodborne pathogens as well as controlling spoilage organisms.

Temperature

Microorganisms are grouped based upon the temperatures at which they grow. Three categories play an important role in foods, psychrotrophs, mesophiles and thermophiles. The storage temperature that certain foods require is driven by what microorganisms are expected to be in a product and if those organisms can increase in number based upon other characteristics such as pH and water activity. Minimum, maximum and optimum temperatures for bacterial foodborne pathogens are shown in Table 1.3. It should be noted that these are approximate values which allow for comparison among foodborne pathogens and should not be taken as absolute.

Psychrotrophs grow at or below 44.6 °F (7 °C) and have optimal growth between 68 °F (20 °C) and 86 °F (30 °C) [4]. Foodborne pathogens that belong to this group are *Listeria monocytogenes*, *Clostridium botulinum* Type E, and *Yersinia enterocolitica*. What should be noted among these microorganisms is their lower temperature limit for growth since, unlike most pathogens, they actively grow in refrigerated conditions, although relatively slowly. Many processors will either include an antimicrobial to restrict growth of these organisms or limit the shelf-life of the food so these organisms cannot grow to sufficient populations to cause illness.

Mesophiles grow well between 68 °F (20 °C) and 113 °F (45 °C) with optima between 86 °F (30 °C) and 104 °F (40 °C) [4]. Bacterial foodborne pathogens that are not psychrotrophs are considered mesophiles. As previously mentioned, given that most of the organisms that make people ill, do so through an infection once consumed, they must thrive well in the range of body temperature. As shown in Table 1.3, examples of organisms in this category include *Campylobacter*, *Escherichia coli*, *Salmonella* spp., *Staphylococcus aureus*, and *Vibrio* spp. Many times, refrigerated or frozen temperatures are selected to store foods to restrict the growth of mesophiles. Examples would be raw beef and chicken which can readily be contaminated with foodborne pathogens such as *Salmonella*. Refrigerated storage will restrict the growth of this organism, although it will not inactivate it, which only occurs when the food is properly cooked.

Thermophiles are the last group and include only spoilage organisms. Thermophiles grow at or above 113 °F (45 °C) with a preferred range of 131–149 °F (55–65 °C) [4]. With respect to foods, these organisms cause spoilage of thermally processed canned foods held at high storage temperatures. These organisms naturally survive the thermal process these products receive and can grow if stored at temperatures in excess of 100 °F (37.7 °C), which can easily occur with outdoor storage during the summer in certain areas.

Table 1.3 Minimum, optimum and maximum temperatures which will support the growth of foodborne pathogens [6]

Organism	Temperature classification	Minimum °F (°C)	Optimum °F (°C)	Maximum °F (°C)
Bacillus cereus	Psychrotroph/Mesophile	41 (5)	82–104 (28–40)	131 (55)
Campylobacter spp.	Mesophile	90 (32)	108–113 (42–45)	113 (45)
Clostridium botulinum types A and B	Mesophile	50–54 (10–12)	86–104 (30–40)	122 (50)
Clostridium botulinum type E	Psychrotroph/Mesophile	37–38 (3–3.3)	77–99 (25–37)	113 (45)
Escherichia coli	Mesophile	45 (7)	95–104 (35–40)	115 (46)
Listeria monocytogenes	Psychrotroph/Mesophile	32 (0)	86–99 (30–37)	113 (45)
Salmonella spp.	Mesophile	41 (5)	95–99 (35–37)	113–117 (45–47)
Staphylococcus aureus	Mesophile	50 (10)	104–113 (40–45)	115 (46)
Shigella spp.	Mesophile	45 (7)	99 (37)	113–117 (45–47)
Vibrio parahaemolyticus	Psychrotroph/Mesophile	41 (5)	99 (37)	109 (43)
Yersinia enterocolitica	Psychrotroph	30 (−1)	82–86 (28–30)	108 (42)

pH

pH is the measurement of acidity or alkalinity of a food. It uses a common scale from 0 to 14, with seven considered to be neutral. Bacteria are similar to most living organisms, in that they prefer a pH around neutrality, with very few growing below 4.0. Minimum pH growth ranges for some foodborne pathogens are shown in Fig. 1.1. While a more neutral pH will allow for faster growth, many foodborne pathogens are capable of growing from 5.0 to 4.0. Spoilage organisms, especially yeast and molds, will be able to grow at even lower pH levels, which is one factor that plays into their ability to act in this role since they can grow in acidic conditions when other competitive microflora cannot.

Of the foods presented in Table 1.4, it can be seen that fruits and some vegetables, such as tomatoes, have natural pH values below that which foodborne pathogens will grow. It should be noted that with every generality there are outliers. Such is the case with melons, which have less acidic pH and will permit the growth of bacteria, including foodborne pathogens. Most of the meats and seafoods have a pH of 5.5 and greater, and as such, pH does not act as a barrier to bacterial growth in these products.

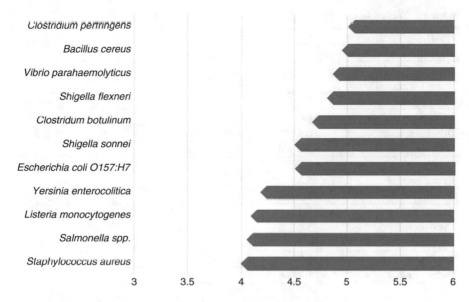

Fig. 1.1 Minimum pH growth ranges for some foodborne pathogens

While some foods are categorized by a natural pH, others have been produced in a manner to achieve a lower pH to help preserve them through inhibition of spoilage and, as an added benefit, pathogenic microorganisms. Fermented products such as kombucha, sauerkraut, kimchi, yogurt, and pickles fall within this category. Organic acids are produced as a by-product of fermentation resulting in a lower overall pH upon completion of fermentation. Other foods are simply mixed with food grade acids, such as acetic acid found in vinegar, to create a quick processed, shelf-stable product. Examples of these foods include salsas, non-fermented pickles and relishes.

Moisture Content

Removal of moisture is one of the oldest methods of preparing shelf-stable foods. Examples of products sold at the farmers market that rely upon low moisture content as their primary means for preservation include breads, beef jerky, dried peppers, dried herbs, and sun-dried tomatoes. Drying of foods helps preserve them through removal of water. When available water is limited, microorganisms such as bacteria, yeast and molds either do not grow or grow very slowly. Food scientists use an uncommon term to the general public to describe the amount of unbound water available for microbial growth known as water activity (a_w). Water activity is technically defined as the ratio of the water vapor pressure of food substrate to the vapor pressure of pure water at the same temperature: $a_w = p/p_0$, where p is the vapor pressure of the solution and p_0 is the vapor pressure of the solvent (usually water).

Table 1.4 Representative pH values of some foods grouped by commodity [4]

Food	pH	Food	pH
Vegetables		*Fruits (cont'd)*	
Asparagus (buds and stalks)	5.7–6.1	Limes	1.8–2.0
Beans (string and Lima)	4.6–6.5	Honeydew	6.3–6.7
Beets (sugar)	4.2–4.4	Oranges	3.6–4.3
Broccoli	6.5	Plums	2.8–4.6
Brussels sprouts	6.3	Watermelons	5.2–5.6
Cabbage (green)	5.4–6.0	*Dairy products*	
Carrots	4.9–5.2; 6.0	Butter	6.1–6.4
Cauliflower	5.6	Buttermilk	4.5
Celery	5.7–6.0	Milk	6.3–6.5
Corn (sweet)	7.3	Cream	6.5
Cucumbers	3.8	Cheese (American mild and cheddar)	4.9; 5.9
Eggplant	4.5		
Lettuce	6.0	*Meat and poultry*	
Onions (red)	5.3–5.8	Beef (ground)	5.1–6.2
Parsley	5.7–6.0	Ham	5.9–6.1
Parsnip	5.3	Veal	6.0
Potatoes (tubers and sweet)	5.3–5.6	Chicken	6.2–6.4
Pumpkin	4.8–5.2	Liver	6.0-6.4
Rhubarb	3.1–3.4	*Fish and shellfish*	
Rutabaga	6.3	Fish (most species)	6.6–6.8
Spinach	5.5–6.0	Clams	6.5
Squash	5.0–5.4	Crabs	7.0
Tomatoes	4.2–4.3	Oysters	4.8–6.3
Turnips	5.2–5.5	Tuna fish	5.2–6.1
Fruits		Shrimp	6.8–7.0
Apples	2.9–3.3	Salmon	6.1–6.3
Apple cider	3.6–3.8	White fish	5.5
Apple juice	3.3–4.1	*Meat and poultry*	
Bananas	4.5–4.7	Beef (ground)	5.1–6.2
Figs	4.6	Ham	5.9–6.1
Grapefruit	3.0	Veal	6.0
Grapes	3.4–4.5	Chicken	6.2–6.4
		Liver	6.0-6.4

Pure water has an a_w of 1.00, and a saturated solution of table salt has an a_w of 0.75. The a_w of most fresh foods is above 0.99. The minimum a_w values reported for the growth of some microorganisms as well as the average a_w of some foods are shown in Table 1.5. Spoilage yeast and molds can grow at the lowest a_w, followed by bacteria, again giving them added leverage in spoilage of food preserved by this method. Among the foodborne pathogens shown in Table 1.5, *Staphylococcus*

Table 1.5 Minimum water activity (a$_w$) values that will allow for growth of microorganisms and approximate a$_w$ values for foods [4, 6, 7]

Organisms	a$_w$	Food	a$_w$
Groups		Fresh fruits, vegetables	0.97–1.00
Most spoilage bacteria	0.90	Fresh meat, poultry, fish	0.99–1.00
Most spoilage yeasts	0.88	Eggs	0.97
Most spoilage molds	0.80	Bread	0.96
		Cheeses	0.95–1.00
Specific organisms		Cured meat	0.87–0.95
Clostridium botulinum, type E	0.97	Maple syrup	0.85
Enterohemorrhagic *Escherichia coli*	0.95	Jellies	0.82–0.94
Salmonella spp.	0.94	Jam	0.80–0.91
Vibrio parahaemolyticus	0.94	Honey	0.75
Clostridium botulinum, types A and B	0.93	Dried fruit	0.60–0.75
Listeria monocytogenes	0.92	Bread crust	0.30
Staphylococcus aureus toxin formation	0.88	Crackers	0.10
Staphylococcus aureus growth	0.83		

aureus can grow at an a$_w$ as low as 0.83, but will only produce toxin, which causes people to become ill if the a$_w$ is at or above 0.88. This is the lowest a$_w$ value for growth of all pathogenic organisms, and for this reason, if dehydration will be the only factor used to make a food shelf-stable, then regulatory guidelines have required the foods' a$_w$ to be at or below 0.85 [5].

It should be understood that while these higher a$_w$ values must be achieved for the organism to grow, many foodborne pathogens have caused illness when consumed in dry foods with a$_w$ values that will not allow for growth. Examples of this include illnesses from dried herbs, nuts, and flour. While these foodborne pathogens cannot multiply in the foods, certain pathogens can survive for long periods of time and can make others ill if not inactivated through a processing step, such as cooking. This is why it is very important to source these dried ingredients from reputable suppliers and always hold dried foods and ingredients so they cannot become contaminated.

Additionally, many foods may rely upon more than one factor to control for the growth of microorganisms. This is referred to as "hurdle technology." For instance, jams and jellies rely on not only the a$_w$ of these products, but also a reduced pH and thermal processing to establish the structure of the products and to eliminate spoilage organisms like molds and yeasts. Very rarely does a single parameter dictate safety, but rather the collective profile of the food. However, understanding the relationships between individual criteria, such as a$_w$, can greatly help with the understanding of why certain criteria are important to the overall safety of a product. As an example, uncontrolled drying in a jerky product could result in too high an a$_w$ resulting in an unsafe food that may support the growth of a foodborne pathogen.

Nutrient Content

Similar to humans, microorganisms must find a source of water, energy, nitrogen, vitamins, and minerals. Water is essential to the growth of microorganisms since metabolic activities occur in an aqueous system inside the organism. Controlling a_w will modulate the available water for microbial growth. Energy can be derived from simple sugars, alcohols and amino acids. A relatively small number of foodborne microorganisms are also able to degrade complex carbohydrates by digesting starches and structural carbohydrates such as lignin and cellulose to simple sugars. Amino acids from proteins are the primary source of nitrogen, with some microorganisms also able to use nucleic acids. Similar to carbohydrate sources, a subset of organisms are also able to digest long and short chain peptides for a nitrogen source. Many organisms are able to synthesize all of the vitamins needed for metabolic function. Those that are limited tend to be deficient with respect to one or more of the vitamin B complexes, and have evolved with mechanisms to secure them from the foods in which they grow. Calcium, iron, magnesium, manganese, sulfur, phosphorus, and potassium are the primary minerals that microorganisms will source from foods that they grow upon. Relatively small amounts of these minerals are required, and some are only required for specific functions rather than growth in general.

While most foods will fulfill the requirements as a substrate for microbial growth given that humans consume them for the same nutritional reasons, it is important to consider adequately removing the nutrient source when cleaning equipment and utensils. Without rigorous cleaning and sanitizing practices, food particulates may remain on equipment and support the growth of microorganisms when not in use. These microorganisms can grow to high numbers if other conditions support growth and act as contaminants when food is prepared subsequently using the equipment or utensils are next used.

Summary

An understanding of the organisms that cause foodborne illness is very helpful when trying to understand the role that food safety best practices play in the foods sold at farmers markets. Ultimately, control strategies are used to inactivate foodborne pathogens if present, restrict them from contaminating a food, and inhibit their growth if they are present. Knowledge of the sources from the environment that may contribute to foodborne pathogen contamination along with the mechanisms that can inhibit bacterial foodborne pathogens from growing in foods can lead to a very good basic understanding of the rationale behind various rules and science-based recommendations used when growing, harvesting and further processing foods.

References

1. Bellemare MF, King RP, Nguyen N (2015) Farmers' markets and food-borne illness. University of Minnesota. http://marcfbellemare.com/wordpress/wp-content/uploads/2015/07/BellemareKingNguyenFarmersMarketsJuly2015.pdf. Accessed 4 Jan 2017
2. Scallan E, Hoekstra RM, Angulo FJ, Tauxe RV, Widdowson MA, Roy SL et al (2011) Foodborne illness acquired in the united states-major pathogens. Emerg Infect Dis 17(1):7–15
3. Kubota K, Kasuga F, Iwasaki E, Inagaki S, Sakurai Y, Komatsu M et al (2011) Estimating the burden of acute gastroenteritis and foodborne illness caused by *Campylobacter*, *Salmonella*, and *Vibrio parahaemolyticus* by using population-based telephone survey data, Miyagi Prefecture, Japan, 2005 to 2006. J Food Prot 74(10):1592–1598
4. Jay JM, Loessner MJ, Golden DA (2005) Modern food microbiology, 7th edn. Springer, New York
5. Acidified Foods (1979) Final rule. Fed Regist 44:16235
6. ICMSF (1996) In: Roberts TA, Baird-Parker AC, Tompkin RB (eds) Microorganisms in foods. Blackie Academic & Professional, London, p 513
7. IFT (2001) Evaluation and definition of potentially hazardous foods. US-FDA. [cited 2016 Dec 29]. Available from: https://www.fda.gov/Food/FoodScienceResearch/SafePracticesforFoodProcesses/ucm094145.htm

Chapter 2
Food Safety Hazards Identified on Small Farms

Judy A. Harrison

Abstract Farmers markets have increased in number in the U.S. by almost 400% since the early 1990s. Customers shop at these markets to get to know the farmers who are producing their food, and to purchase products they view as more nutritious, better tasting, higher quality, better for the environment and safer than foods from larger, commercial farms being sold in supermarkets. Yet studies in the U.S. and in other countries have identified food safety hazards on farms and in farmers markets that may increase the risk of foodborne illnesses. Risky practices on farms include the use of raw manure without appropriate waiting periods observed between application and harvest, use of untested well or surface water for irrigation and/or washing of produce, lack of sanitary facilities and handwashing facilities for workers, lack of training for workers, food contact surfaces not properly cleaned and sanitized and lack of temperature control both on the farm and during transport to market. Hazards have also been identified with livestock and poultry products such as lack of sanitation and temperature control. A lack of sanitation practices and microbial problems associated with the use of raw milk have been identified as hazards on farms making and selling artisanal cheeses.

Keywords Farm food safety • Farmers market food safety • Food safety risks • Food safety hazards

The numbers of farmers markets have increased in the U.S. by almost 400% since the early 1990s signifying the increasing popularity of the local food movement (Fig. 2.1). This increase in visibility and popularity has been fueled by campaigns such as *Know Your Farmer, Know Your Food* and *The People's Garden* [1].

Several studies have examined the reasons why consumers shop at farmers markets. Reasons include beliefs that purchasing from the farmers market supports

J.A. Harrison (✉)
Department of Foods and Nutrition, University of Georgia,
204 Hoke Smith Annex, Athens, GA 30602, USA
e-mail: judyh@uga.edu

© Springer International Publishing AG 2017   13
J.A. Harrison (ed.), *Food Safety for Farmers Markets: A Guide to Enhancing Safety of Local Foods*, Food Microbiology and Food Safety,
DOI 10.1007/978-3-319-66689-1_2

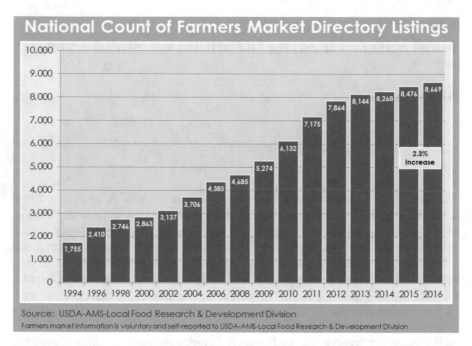

Fig. 2.1 National count of farmers market directory listings. Reprinted from http://www.ams. usda.gov, by AMS-USDA, Transportation and Marketing Program, Local Food Research and Development Division. Available from:https://www.ams.usda.gov/sites/default/files/media/ National%20Count%20of%20Operating%20Farmers%20Markets%201994-2016.jpg

local farmers and is better for the environment, the food is fresher with better quality and flavor, the food is a better value for the money, the food is more likely to be grown locally and is more traceable than products from the grocery store, and the food is safer when produced locally [2–4].

In 2008, 81% of farms selling locally were small farms with less than $50,000 in gross sales, and 14% were medium sized farms selling between $50,000 and $250,000 in gross sales [5]. Studies have shown that often farmers selling at farmers markets are relatively new to farming and have less experience than those on larger, more established farms [6, 7]. According to Martinez et al. [8], the 2007 U.S. Census of Agriculture indicated that farmers selling directly to consumers had four years less experience than those not marketing directly to consumers, and 40% were beginning farmers with less than 10 years of experience. The 2012 U.S. Census of Agriculture identified little change in most categories compared to the 2007 census [9]. Harrison et al found that out of 328 participants in small farm produce safety trainings in Georgia, South Carolina and Virginia during the period from 2011 to 2013, 43% indicated farming for less than 3 years, 20% from 4 to 9 years and 36% for 10 or more years. Of those participants, 34% used organic methods, 29% used conventional methods and 37% used both organic and conventional farming methods [10]. Laury-Shaw et al.

noted that prior to an initial food safety training, less than 10% of participants had food safety plans that included written policies regarding worker attire; worker behaviors involving eating, drinking or smoking while working with products and handling during transportation [7]. Parker et al. [11] stated that there is little knowledge of how food safety is handled on small and medium farms, even though a study 6 years earlier in 2006 by Simonne et al. found that out of 47 farmers market vendors, 50% thought food safety was very important and were very confident of their food safety practices, but only 32% had completed any type of food safety training [12]. The study by Parker et al. about food safety concerns among growers found that regardless of the size of farm, growers were most concerned about consumer behavior and health and hygiene of workers [11]. Other lesser concerns for large growers included sanitation of facilities and equipment, wildlife fecal contamination and quality of water for irrigating and washing produce. Lesser concerns for medium growers were similar with these growers including pesticide application and soil amendments as well. Small growers in the study included the presence of wildlife feces and pesticide drift as concerns, but only a few small growers (≤ 18) included concern over sanitation of facilities and equipment, manure use and water quality even though these issues are of concern to food safety experts and are addressed as part of good agricultural practices.

Although these studies indicate some level of awareness among local food producers of potential issues and conditions that could affect the safety of produce grown on small to medium farms and sold in farmers markets, a multi-state survey identified growing and handling practices on small to medium sized farms that could put consumers at risk for foodborne illnesses [6]. Out of 226 farmers responding to a survey, 128 (57%) used manure with 18% of those using a mixture of raw and composted manure (including one report of using humanure, human manure from a composting toilet). Almost 15% of manure users applied manure to fields more than twice a year, raising concerns that the recommended 90 day and 120 day waiting periods between application of raw manure and harvest for crops that do not touch the soil and crops that touch the soil, respectively, that are recommended in the National Organic Program are not being met [13]. Although most growers used tested water sources for irrigation, 30.5 % of respondents used untested well water or rainwater and surface water from streams or ponds, for irrigation which has the potential for microbial contamination. In terms of worker hygiene, approximately 66% of respondents reported having sanitary facilities and handwashing facilities available near fields and packing sheds. Yet the lack of facilities at many farms makes hand hygiene questionable and raises concern about potential contamination. Fifty percent of the operations indicated that crops are harvested with bare hands. Only 41% of the farmers indicated they had offered sanitation training to their workers [6].

Harrison et al. also identified post-harvest handling practices that could increase a consumer's risk for foodborne illness [6]. Approximately 16% of the farmers who responded used untested well water, surface water (such as ponds, streams or springs) and rainwater for washing produce after harvest. Only 39% of respondents sanitized surfaces that come in contact with produce at the farm, and only 33%

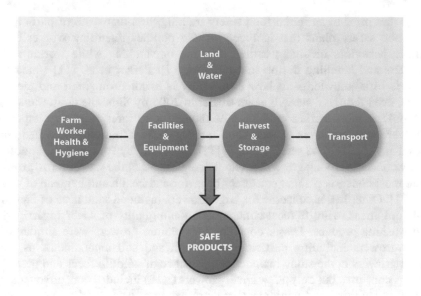

Fig. 2.2 Model of Food Safety on the farm showing areas where best practices must be implemented for production of safe produce destined for farmers markets. Reprinted from Enhancing the Safety of Locally Grown Produce—On the Farm. Harrison JA. 2012. University of Georgia Cooperative Extension Publication #FDNS-E-168-2

always cleaned containers used to transport produce to market between uses. Cooling of produce on the farm or during transport was also lacking with 18.1% reporting no cooling methods used on the farm and 35% rarely or never cooling produce during transport to market. These findings indicate the potential for food safety hazards associated with growing, harvesting and post-harvest handling of produce destined for the farmers market and other direct-to-consumer outlets. Based on these findings, a model illustrating areas where best practices must be used to enhance produce safety has been developed and is presented in Fig. 2.2. An assessment tool is presented in Fig. 2.3 for use on small farms selling direct market produce. A detailed description of best practices to control food safety hazards for produce grower/vendors is presented in Chap. 4.

Although produce farms accounted for more than half of direct sales to consumers in the 2007 U.S. Census of Agriculture, 7% of livestock producers also sold directly to consumers with beef, poultry, dairy and eggs accounting for the highest percentages of products sold [8]. According to Painter et al., meat and poultry (beef, game, pork and poultry), accounted for fewer illnesses than produce in the period from 1998 to 2008 but accounted for a higher percentage of deaths due to foodborne illness (29% versus 23%) [14]. While the debate continues over issues related to animal welfare, use of antimicrobials, etc. and the safety of meat products from "industrial farm animal production" systems versus small-farm production, foodborne illness can be associated with either system [15]. In 2010, the number of slaughter facilities in the U.S. had decreased

Farm Self-Help Form

$\mathcal{P}$ractice	YES	NO	DOES NOT APPLY TO ME
Training & Certifications			
Our farm has established food safety rules and practices.			
Our farm has completed food safety trainings and/or certification courses.			
Our farm has records of certification or evidence of training to help ensure food safety.			
Land & Water Use			
I know the land use history, whether the farm was previously used for livestock production or has a history of application of biosolids, septage or other by-products containing feces.			
My crop production areas are separate from or NOT located near dairy, livestock or poultry production areas or where run-off from such areas could be possible.			
If crop production areas are near or adjacent to dairy, livestock or poultry production areas, I make sure natural or physical barriers will prevent contamination of the produce growing area by wind or water.			
If I use raw animal manure, I wait at least 120 days between application and harvest for crops touching the soil and 90 days for other crops.			
I NEVER use septage or untreated human manure in crop production.			
Any composted manure I use follows the U.S. EPA or National Organic Program recommendations for temperature, turning and time to reduce disease-causing microorganisms.			
I have my well water that I use for irrigation tested for the presence of bacteria.			
I NEVER use untreated surface water (ponds, lakes, streams or springs) for overhead irrigation.			
I use municipal water or tested well water for overhead irrigation.			
I have my well water that I use for rinsing fruits and vegetables tested for the presence of bacteria.			
I NEVER use surface water (ponds, lakes, streams or springs) for rinsing fruits and vegetables.			
Farm Worker Hygiene			
I have policies in place to limit sick workers from coming in contact with fruits and vegetables.			
I provide sanitation training for my workers.			
I provide training for my workers on proper glove use.			
My workers have access to handwashing facilities with clean water, soap and paper towels within a short walking distance of my fields.			

Fig. 2.3 Enhancing the safety of locally grown produce—farm self-help form. Courtesy of Judy Harrison. University of Georgia Publication #FDNS-E-168-1

from over 1200 in the 1990s to around 800 with the consolidation of the meat industry [16]. This closure of many facilities has made it necessary for livestock producers wanting to sell products to seek alternative methods of slaughter and packing, either small state inspected facilities willing to serve small businesses or using the services of mobile abattoirs [17]. Regardless of the situation or

$\mathcal{P}\text{ractice}$	YES	NO	DOES NOT APPLY TO ME
My workers have access to toilet facilities within a short walking distance of my packing areas.			
I train my workers to seek immediate first aid for injuries like cuts, abrasions, etc. that could be a source of contamination for produce.			
I have trained my workers on what to do with produce that comes in contact with blood or other bodily fluids.			
Facilities & Equipment			
Toilet facilities are serviced and cleaned on a regular schedule.			
Handwashing facilities are cleaned and stocked with clean water, soap and paper towels on a regular schedule.			
Harvesting equipment (knives, pruners, machetes, etc.) is kept reasonably clean and is sanitized on a regular basis.			
Harvesting containers and hauling equipment are cleaned and/or sanitized between uses.			
Surfaces that come in contact with fruits and vegetables at my farm are cleaned and sanitized regularly.			
Damaged containers are properly repaired or discarded.			
Any cardboard boxes used are new and only used once.			
Storage & Transport			
Produce is handled carefully and packed securely to prevent bruising and injury.			
I cool fruits and vegetables after harvest.			
Produce is kept cool during transport to market.			
Containers used with fruits and vegetables are cleaned and sanitized between each use.			
The vehicle is NOT used to transport animals, raw manure, chemicals or any other potential contaminant.			
The vehicle used to transport fruits and vegetables is cleaned frequently.			

> If you answered "no" to any of the questions, those questions represent areas where changes or improvements may help your farm to offer safer products, attract more customers because of your commitment to food safety and reduce potential risk of foodborne illness. Please read the *Enhancing the Safety of Locally Grown Produce* factsheets for your risk area to learn how to minimize risk.

This project was supported all, or in part, by a grant from the National Institute of Food and Agriculture, United States Department of Agriculture (Award Number 2009-51110-20161).

Publication #FDNS-E-168-1. J.A. Harrison, J.W. Gaskin, M.A. Harrison, J. Cannon, R. Boyer and G. Zehnder.   February 2012

The University of Georgia and Ft. Valley State University, the U.S. Department of Agriculture and counties of the state cooperating. Cooperative Extension, the University of Georgia Colleges of Agricultural and Environmental Sciences and Family and Consumer Sciences, offers educational programs, assistance and materials to all people without regard to race, color, national origin, age, gender or disability. An Equal Opportunity Employer/Affirmative Action Organization, Committed to a Diverse Work Force.

Fig. 2.3 (Continued)

method used, adherence to strict sanitation practices, cooling practices and time-temperature control would be essential to minimize foodborne illness risks. USDA's Food Safety and Inspection Service provides guidance documents for mobile processing units [18].

A study of poultry products sold at farmers markets versus those conventionally processed and sold at supermarkets in Pennsylvania identified significantly higher levels of generic *E. coli*, total coliforms, *Salmonella* spp. and *Campylobacter* spp. in whole chicken sold at farmers markets [19]. This study found increased risk of food safety hazards associated with poultry sold in farmers markets. A detailed description of food safety considerations for meat and poultry vendors at farmers markets is presented in Chap. 5.

In addition to meat and poultry, dairy products are also produced on small farms and sold at farmers markets. Although some states prohibit the sale of raw milk, other states allow the sale which can also increase food safety risks for consumers when sold through farm stands or farmers markets. Raw milk has a historic association with foodborne illness due to the presence of foodborne pathogens. Painter et al. noted a higher incidence of *Campylobacter* associated with raw milk [14]. From 2007 to 2012, 26 states reported 81 outbreaks to the Centers for Disease Control and Prevention (CDC) caused by raw milk, an increase from 30 outbreaks from 2007 to 2009 to 51 outbreaks between 2010 and 2012 [20]. The hazards associated with raw milk outbreaks in this report were *Campylobacter* (81% of outbreaks), shiga toxin-producing *E. coli* (17% of outbreaks) and *Salmonella* (3% of outbreaks). However, outbreaks have included a multistate outbreak of listeriosis linked to raw milk from an organic producer in Pennsylvania [21].

Many small dairy farms also make and sell artisan cheeses at farmers markets and other venues. In addition to risks from environmental conditions and sanitation issues on farms, some farmstead cheesemakers use raw milk as an ingredient and rely on proper aging to eliminate pathogens [22]. Outbreaks of foodborne illnesses linked to raw milk and raw milk cheeses have raised concerns about the safety of these products. Regulations for the production of raw milk cheeses require that cheeses be aged for not less than 60 days at a temperature of not less than 35 °F (2 °C) [23]. However, studies of artisan and farmstead cheesemakers historically have identified varying levels of risk associated with these products. In a study of 11 cheesemaking facilities, D'Amico et al. reported that 8 of the 11 facilities (73%) had milk samples that tested positive for *Staphylococcus aureus* (46 of 133 samples or 34.6 %), three milk samples (2.3%) tested positive for *Listeria monocytogenes* and one for *Escherichia coli* O157:H7 [24]. *Salmonella* was not found in any of the samples [24]. Another study by D'Amico et al. indicated that if contamination with *L. monocytogenes* occurs during the post-processing period of soft, mold-ripened cheeses, the 60-day aging period may not be adequate to ensure safety [25]. Machado et al. [22] reported that observations of five farmstead cheesemaking facilities in Pennsylvania and survey responses of state inspectors indicated that improvements were needed in basic sanitation, although cheesemakers had rated their knowledge of food safety, their attitudes toward food safety and their handling practices as good to very good. Outbreaks of illness have been associated with farmstead cheesemaking facilities. Evidence prompted a recall of 14 cheese varieties potentially linked to a *Salmonella* outbreak that sickened 100 people in 2016 from a farmstead creamery in North Carolina which had sold cheese through retail locations, farmers markets and restaurants throughout North Carolina, Tennessee, South Carolina, Virginia and Georgia [26].

Safety of food in farmers markets, as well as any other venues where food is sold, requires strict attention to good agricultural practices, good manufacturing practices and proper sanitation on farms where the food is produced. Improper handling on the farm can lead to increased risk of foodborne illnesses from farmers markets.

Summary

Studies have identified the potential for food safety hazards to exist on small farms selling products directly to consumers through farmers markets and other venues. Self-reported data as well as direct observations have noted problems with hand hygiene, sanitation and temperature control on farms. These conditions could lead to an increased risk of contamination of products and foodborne illnesses among consumers.

References

1. U.S. Dept. of Agriculture, Center for Nutrition Policy and Promotion. Know your farmer, know your food – growing a healthier you [cited 2017 Apr 6]. Available from: https://www.cnpp.usda.gov/KnowYourFarmer
2. Worsfold D, Worsfold PM, Griffith CJ (2004) An assessment of food hygiene and safety at farmers' markets. Int J Environ Health Res 14(2):109–119
3. Crandall PG, Friedly EC, Patton M, O'Bryan CA, Gurubaramurugeshan A, Seideman S, Ricke SC, Rainey R (2011) Consumer awareness of and concerns about food safety at three Arkansas farmers' markets. Food Prot Trends 31(3):156–165
4. Wolf MM, Spittler A, Ahern J (2005) A profile of farmers' market consumers and the perceived advantages of produce sold at farmers' markets. J Food Distrib Res 36(1):192–201
5. Low SA, Vogel S (2011) Direct and intermediated marketing of local foods in the United States, ERR-128, U.S. Department of Agriculture, Economic Research Service. Nov [cited 2016 Sept 16]. Available from: https://www.ers.usda.gov/webdocs/publications/err128/8276_err128_2_.pdf
6. Harrison JA, Gaskin JW, Harrison MA, Cannon JL, Boyer RR, Zehnder GW (2013) Survey of food safety practices on small to medium-sized farms and in farmers' markets. J Food Prot 76(11):1989–1993
7. Laury-Shaw A, Strohbehn C, Naeve L, Wilson L, Domoto P (2015) Current trends in food safety practices for small-scale growers in the midwest. Food Prot Trends 35(6):461–469
8. Martinez S, Hand M, DaPra M, Pollack S, Ralston K, Smith T, Vogel S, Clark S, Lohr L, Low L, Newman C (2010) Local foods systems: concepts, impacts and issues. USDA ERS Report No. 97 [cited 2016 Aug 26]. Available from: https://ideas.repec.org/p/pra/mprapa/24313.html
9. U.S. Department of Agriculture National Agricultural Statistics Service (2012) Census of agriculture. 2014 Aug [cited 2016 Aug 26]. Available from: https://www.agcensus.usda.gov/Publications/2012/Full_Report/Volume_1,_Chapter_1_US/usv1.pdf
10. Harrison JA, Gaskin JW, Harrison MA, Cannon JL, Boyer RR, Zehnder GW (2013) Enhancing the safety of locally grown produce through extension education for farmers and market managers [abstract]. J Food Prot 76(Suppl A):P1–87
11. Parker JS, Wilson RS, LeJeune JT, Doohan D (2012) Including growers in the "food safety" conversation: enhancing the design and implementation of food safety programming based

on farm and marketing needs of fresh fruit and vegetable producers. Agric Hum Values 29:303–319

12. Simonne A, Swisher M, Saunders-Ferguson K (2006) Food safety practices of vendors at farmers' markets in Florida. Food Prot Trends 26(6):386–392

13. Code of Federal Regulations (2017) National Organic Program, Title 7, Subtitle B, Chapter 1, Subchapter M, Part 205, Section 205.203

14. Painter JA, Hoekstra RM, Ayers T, Tauxe RV, Braden CR, Angulo FJ, Griffin PM (2013) Attribution of foodborne illnesses, hospitalizations, and deaths to food commodities by using outbreak data, United States, 1998–2008. Emerg Infect Dis 19(3):407–415

15. Rossi J, Garner SA (2014) Industrial farm animal production: a comprehensive moral critique. J Agric Environ Ethics 27:479–522

16. Johnson RJ, Marti DL, Gwin L (2012) Slaughter and processing options and issues for locally sourced meat. USDA ERS Report No. LDP-M-216-01. June [cited 2016 Sept 16]. Available from: https://www.ers.usda.gov/publications/pub-details/?pubid=37460

17. Thompson S (2010) Going mobile – co-ops operate traveling slaughter units to help grow local foods movement. USDA Rural Development. Rural Cooperatives 77(6):4–7. Available from: http://www.rd.usda.gov/files/CoopMag-nov10.pdf

18. U. S. Department of Agriculture Food Safety and Inspection Service (2010). Mobile Slaughter Unit Compliance Guide [cited 2016 Sept 16]. Available from: https://www.fsis.usda.gov/shared/PDF/Compliance_Guide_Mobile_Slaughter.pdf

19. Scheinberg J, Doores S, Cutter CN (2013) A microbiological comparison of poultry products obtained from farmers' markets and supermarkets in Pennsylvania. J Food Saf 33:259–264

20. Mungai EA, Behravesh CB, Gould LH (2015) Increased outbreaks associated with nonpasteurized milk, United States, 2007–2012. Emerg Infect Dis 21(1):119–122

21. Centers for Disease Control and Prevention (2016) Multistate outbreak of listeriosis linked to raw milk produced by Miller's Organic Farm in Pennsylvania (Final Update) [cited 2016 Dec 28]. Available from: http://www.cdc.gov/listeria/outbreaks/raw-milk-03-16/index.html

22. Machado RAM, Radhakrishna R, Cutter CN (2017) Food safety of farmstead cheese processors in Pennsylvania: an initial needs assessment. Food Prot Trends 37(2):88–98

23. Code of Federal Regulations (2017) Cheese from unpasteurized milk, Title 7, Subtitle B, Chapter 1, Subchapter C, Part 58, Subpart B, Section 58.439

24. D'Amico D, Groves E, Donnelly CW (2008) Low incidence of foodborne pathogens of concern in raw milk utilized for farmstead cheese production. J Food Prot 71(8):1580–1589

25. D'Amico D, Druart M, Donnelly CW (2008) 60-Day aging requirement does not ensure the safety of surface-mold-ripened soft cheeses manufactured from raw or pasteurized milk when *Listeria monocytogenes* is introduced as a post-processing contaminant. J Food Prot 71(8):1563–1571

26. U. S. Food and Drug Administration (2016) Chapel Hill creamery recalls cheese products because of possible health risk [cited 2016 Dec 28]. Available from: https://www.fda.gov/Safety/Recalls/ucm513946.htm

Chapter 3
Potential Food Safety Hazards in Farmers Markets

Judy A. Harrison

Abstract Consumers often view foods sold at farmers markets as healthier and safer than foods produced and shipped long distances and sold in stores. However, outbreaks of foodborne illnesses have occurred from products sold at farmers markets. Studies in the U.S. have shown that rules regarding farmers markets and the food products allowed for sale vary from state to state. A survey of farmers market managers found that few have written food safety plans for their market, and many managers do not ask questions of growers or vendors as to how products are produced or processed. Products in U.S. markets include fresh produce and products made in small businesses, many of which may be operated in home kitchens under cottage food regulations. Studies indicate that threats to the safety of products sold in farmers markets can be categorized into threats from the environment, threats due to the infrastructure or facilities and threats from people—both vendors and customers. Observations have included a lack of handwashing and sanitation, lack of refrigeration, pets in the market, vendors eating and drinking while handling foods and other issues that could put customers at risk. Regulatory personnel and Extension food safety educators responding to a survey indicated that it is prevalent to very prevalent for owner/operators (55–74%) with whom they work to view their products as unlikely to cause illness because their businesses are small, local or organic. However, respondents noted a lack of food safety knowledge, including knowledge of allergens, required labeling and other regulatory requirements among these vendors. These and other practices could put customers in farmers markets at greater risk for foodborne illnesses.

Keywords Farmers market food safety • Cottage foods • Very small and small food businesses

J.A. Harrison (✉)
Department of Foods and Nutrition, University of Georgia,
204 Hoke Smith Annex, Athens, GA 30602, USA
e-mail: judyh@uga.edu

© Springer International Publishing AG 2017
J.A. Harrison (ed.), *Food Safety for Farmers Markets: A Guide to Enhancing Safety of Local Foods*, Food Microbiology and Food Safety,
DOI 10.1007/978-3-319-66689-1_3

Although products sold at farmers markets are often viewed by customers as being healthier and safer for them than food sold in supermarkets, there have been outbreaks of illness from products sold at farmers markets and/or pathogens identified in sampled products. For example, one case of trichinellosis from wild boar meat sold at a farmers market was reported [1]. An outbreak caused by *Salmonella* linked to Mexican food products sold at farmers markets in Iowa resulted in 44 illnesses and five hospitalizations in 2010 [2]. In 2011, strawberries sold at roadside stands and farmers markets were linked to an outbreak of *E. coli* O157:H7 which resulted in one death and approximately 15 illnesses [3]. In 2012, cheese sold at farmers markets in Washington State was recalled when a sample of the product taken by the state department of agriculture tested positive for *Listeria*, however no cases of illness were reported [4]. *Campylobacter* from raw milk sold at an on-farm store in Pennsylvania and off-site resulted in approximately 148 cases of illness [5]. In 2013, an outbreak caused by *Salmonella* linked to pasteurized cashew cheese sold in health food stores and in a farmers market in California sickened at least 15 people in the western U.S. [6]. Improperly processed jarred pine nut basil pesto sold from a farm stand in California in 2014 was responsible for two cases of botulism and a warning issued that "Consumers at farm stands and markets should be aware of the risk from improperly canned foods, including those in jars, produced without licensure and oversight from regulatory bodies" [7]. In a study of herbs sold at farmers markets in California and Washington State, researchers tested 133 samples of cilantro, basil and parsley from 13 different farmers markets and found that 24.1% of samples tested positive for generic *E. coli* which could be indicative of a lack of sanitation, and one sample tested positive for *Salmonella* [8].

Simonne et al. noted in a 2006 study that items being sold in markets in Florida were predominantly vegetables, fruits, herbs and flowers, with a few vendors selling meats, eggs and value-added products [9]. Subsequently, nationwide findings in the 2012 USDA Census of Agriculture indicated that almost 95,000 farms in the U.S. produced and sold value added products. These included items such as jerky, jams, jellies, preserves, cider and wine. This was most prevalent in Texas, California, Kentucky, Missouri and Oklahoma [10]. Although fresh, whole produce is widely sold in farmers markets, observations of products have included breads (Fig. 3.1); cheeses; pastas (Fig. 3.2); jams and jellies (Fig. 3.3); microgreens (Fig. 3.4); mushrooms (Fig. 3.5); granola and nuts (Fig. 3.6); coffee; and juices or smoothies that are blended from vegetables and fruits sold in the market, in some instances, with no method for thorough washing of juicers or blender jars between uses for the duration of the market (Fig. 3.7). Observations have included further processed produce such as diced onions, shredded lettuce, etc. being sold in markets with no refrigeration (Fig. 3.8). Some markets also have included an assortment of crafts, soaps, lotions and health-promoting "tonics" (Fig. 3.9) [11]. In addition to selling through farmers markets, almost 13,000 farms in 2012 participated in community supported agriculture arrangements (CSAs) in which produce and other products are delivered to subscribers thus increasing access to these local products [10].

Studies have shown that rules and regulations regarding farmers markets and the food products allowed for sale in these markets are not consistent throughout the U.S. A 2013 study by the Harvard Food Law and Policy Clinic found that 42

Fig. 3.1 Display of breads without packaging or protection from contamination in a farmers market. Courtesy of Judy Harrison

Fig. 3.2 Pastas for sale in a farmers market. Courtesy of Judy Harrison

states had cottage food laws, and that these regulations not only varied from state to state, but also were in some cases, difficult to find on state government websites or were not clearly defined [12]. The study found that even the types of products allowed to be made in home kitchens and to be sold as cottage foods varied from state to state, even though most states specify that only non-potentially hazardous

Fig. 3.3 Jams and jellies for sale in a farmers market. Courtesy of Judy Harrison

Fig. 3.4 Microgreens for sale in a farmers market. Courtesy of Judy Harrison

foods (foods that do not require time-temperature control for safety) can be sold in this way. The study also found wide variations from state to state regarding limits on the amount of products that can be sold under cottage food regulations; requirements for special licenses or permits; requirements for inspections; and application fees for licensing.

In an effort to determine more about the training needs of these small food producers, Harrison et al. surveyed a convenience sample of state food safety regulatory officials and food safety educators nationwide to determine their

Fig. 3.5 Mushrooms for sale in a farmers market. Courtesy of Judy Harrison

Fig. 3.6 Granola, nuts and mixes for sale in a farmers market. Courtesy of Judy Harrison

observations and perceptions of the prevalence of practices that could contribute to food safety risks in food products from small and very small businesses and to determine their observations and opinions of the prevalence of certain food safety risks in farmers markets [13]. The study found that out of 65 respondents to the survey, 70% indicated that cottage food businesses were allowed in their

Fig. 3.7 Blender station for making smoothies from fruits and vegetables sold in a farmers market without a method to thoroughly clean jars between uses. Courtesy of Judy Harrison

state. Survey participants indicated a prevalence rating of only somewhat prevalent (representing a range of 25–39% of owner/operators) for owner/operators having a good understanding of good agricultural practices, good manufacturing practices, and food safety as it relates to their products and the processes they use to produce them. Also rated as only somewhat prevalent were owners/operators having knowledge of labeling requirements for their products. Specifically related to allergen labeling, only 10–24% of owner/operators were rated as being able to identify all eight major food allergens and mandatory labeling requirements representing a rating of slightly prevalent. Survey participants indicated that it was prevalent for owners/operators of small and very small businesses (55–74% of owner/operators) to view their products as unlikely to cause illness because their businesses are small, local or organic. This raises an important issue as it relates to adoption of food safety practices. If processors/vendors think their products are unlikely to cause illness, then there may be little or no motivation to change their food handling behaviors. The participants in this survey also indicated a slightly prevalent rating for vendors at farmers markets selling low-acid canned foods without appropriate licenses (10–24% of vendors). This raises concern about the risk of botulism in home-canned products sold in farmers markets [13].

Fig. 3.8 Diced onions and other further processed vegetables for sale without refrigeration in a farmers market. Courtesy of Judy Harrison

A more detailed discussion of regulations related to manufactured food products sold in farmers markets is presented in Chap. 6.

Harrison et al. developed a model of food safety for farmers markets regarding the sale of fresh produce (Fig. 3.10). This model is indicative that in addition to production and handling of products on the farm and during transport to market, conditions in farmers markets and practices of both market vendors and customers can impact the safety of products sold in the markets. Many aspects of the model can also apply to other products besides produce.

Unique conditions associated with farmers markets that increase opportunities for contamination have been noted in several studies and can be categorized as threats to product safety. As noted by Worsfold et al. in a study of farmers markets in the U.K. published in 2004, conditions at these markets are often somewhat lacking with vendors selling products outdoors with foods exposed to environmental contaminants and markets having little access to potable water for handwashing or washing of products and little or no electricity for refrigeration [14]. Conditions in many U.S. markets are similar. Data from a survey of farmers market managers published in 2013 indicated that over 40% of the 45 respondents (18 of 45) reported having no food safety plan in place for their market. Over 80% of the managers (36 of 45) indicated they do not ask any questions to farmers about practices on the farm that could affect the safety of the produce they grow such as using animal manure, allowing domestic animals in production areas, providing bathroom and handwashing facilities near fields and packing sheds, using sanitizers on food contact surfaces and posting signs or teaching workers about sanitation procedures. Responses indicated

Fig. 3.9 Juice drinks and
"health promoting"
cleanses available for sale
in a farmers market.
Courtesy of Judy Harrison

a lack of sanitation in the market with only 11% of the managers (5 of 45) indicating they always clean market containers between uses, and less than 25% (11 of 45) indicating they sanitize surfaces at the market. Over 75% of these managers offered no sanitation training for their market workers or the vendors at their markets, even though on-site food preparation was allowed by 27% of the managers. In addition, handling practices such as cooling were lacking in the markets with more than 50% of the managers reporting that no cooling methods are used [15]. Although this data was self-reported by managers rather than gathered through actual observations of practices, it did provide an early indication of potential food safety hazards that may be associated with products sold in farmers markets.

Threats to the safety of products sold in farmers markets can be categorized into threats from the environment, threats due to the infrastructure or facilities and from people—both vendors and customers [14]. Unlike grocery stores, many farmers

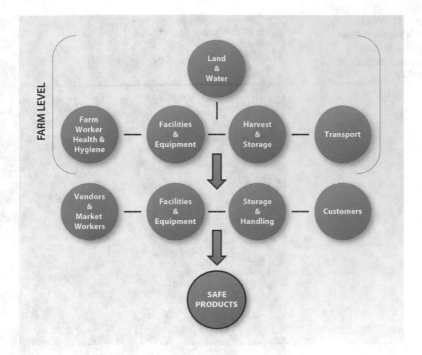

Fig. 3.10 Model of food safety at the market showing areas where best practices must be implemented for product safety both at the farm and in the farmers market. Courtesy of Judy Harrison. Reprinted from Enhancing the Safety of Locally Grown Produce—at the Market. Harrison JA, Gaskin JW, Harrison MA, Cannon J, Boyer RR, Zehnder GW. 2012. University of Georgia Cooperative Extension Publication #FDNS-E-168-12

markets are housed in outdoor environments. Environmental threats such as unclean surfaces for product storage and display (Fig. 3.11), foods stored or displayed directly on the ground (Fig. 3.12), presence of animals including pets (Fig. 3.13), and/or bird droppings, insects, and in some cases, live animals being sold in the midst of produce displays (Fig. 3.14) can contribute to the risk of contamination. Customers can be observed touching animals and then continuing to shop and touch food products without handwashing or use of hand sanitizer.

Another category of threats is related to infrastructure or facilities. As noted in a study by Harrison et al., many market managers lack food safety plans or rules for their markets [15]. Markets may be in locations that lack electricity for cooling products and keeping hot foods hot. Some markets lack appropriate handwashing and bathroom facilities for workers and for customers, as well as, water for cleaning. Poor infrastructure such as this can contribute to a third category of threats, threats due to people [14]. Studies in farmers markets have observed little, if any, handwashing before handling foods; vendors eating and drinking while handling foods; vendors and/or customers handling pets or other animals in the market and not washing hands before touching food products; vendors using gloves incorrectly (touching clothing, body, trash, money, etc.) and then touching foods; lack of appropriate utensils or thorough washing of utensils and equipment between uses

Fig. 3.11 Produce displayed on unclean surfaces in farmers markets. Courtesy of Judy Harrison

(e.g. during cutting of samples with pocket knives, blending of smoothies containing fruit and vegetables purchased at the market without cleaning blenders between uses, etc.) [16, 17]. Behnke et al. observed 18 farmers market workers in Indiana preparing and serving food on-site. The study found hand hygiene to be practically non-existent with only two attempts to wash hands observed in 417 transactions where handwashing would have been in compliance with state health department guidelines [16]. McIntyre et al. observed seven market sites in British Columbia. Their study indicated that over 50% of vendors (11 of 21) stored food boxes directly on the ground, and 43% did not pre-package food or use sneeze guards to protect uncovered foods. Four out of 11 vendors who provided samples for tasting (36.4%) did not have required handwashing stations at their stalls. Two vendors had the stations but did not have appropriate handwashing supplies. Most vendors (91%) handled both money and food. Even when gloves were used, they were used inappropriately [17].

Some studies have observed little, if any, use of thermometers to check cold or hot holding temperatures or temperatures of displays of higher risk produce and have observed samples of cut fruits and vegetables without temperature control [18]. In addition, temperature control was lacking in the markets observed in the McIntyre et al. study [17] with violations observed involving vendors displaying

eggs out of refrigeration, vendors selling baked goods with whipped cream toppings at ambient temperature and frozen fish being sold from an open cooler with visible thawing. Only one of ten vendors with products requiring refrigeration in the study had a thermometer to check the temperature of their cooling equipment.

Vandeputte et al. observed food handling practices of a sample of 26 vendors from 14 farmers markets in Rhode Island using a smartphone application to conceal direct observation [18]. Ten of the 14 markets (71%) were described as unclean due to environmental contaminants such as geese droppings and the presence of pets in the market. Pets are a common occurrence in farmers markets. In 14 markets observed in the U.S., France and Spain, 64% (9 of 14) had pets, some on leashes, some not, even with "no pets allowed" signs posted in at least three of the markets with pets [11]. Even when pets are on leashes, they can be observed coming in contact with produce in displays that are low to the ground when owners are distracted, and owners and shoppers can be observed handling the animals and then touching food items that are for sale, providing opportunities for contamination of products being sold in the markets [11].

Fig. 3.13 Pet and owner
interaction in a farmers
market. Courtesy of Judy
Harrison

In addition to environmental contaminants from animal presence, Vandeputte et al. categorized 73% of vendor stands (19 of 26) as "dirty" due to produce tables that were visibly soiled, dirty water in containers and/or visible dirt or holes in the tent coverings for the stalls. No vendors were observed washing their hands, even though 81% touched money, and 32% ate and drank as they handled produce. No handwashing facilities were available at the stands [18].

An observational study of farmers markets in Virginia using a secret shopper method via a smartphone application found no difference in the prevalence of risky food handling practices between produce vendors in three markets who had attended a food safety training program and vendors in two markets who had not been trained [19]. Of the 42 produce vendors observed, 28.1% of those trained (9 of 32) still stored produce at ground level in cardboard boxes as did 10% (one of 10) of the untrained vendors. Vendors also displayed food on surfaces made of wood that could not be properly cleaned and sanitized (37.5% of trained; 40% of untrained) and/or used wood or re-used cardboard containers that could not be properly cleaned and sanitized for holding produce (46.9% of trained; 40 % of untrained). For wood to be used with produce, the U.S. Food and Drug Administration (FDA) indicates that it should be washed, sanitized and allowed to completely air-dry between uses. This, coupled with the porous nature of wood make it difficult, if not

Fig. 3.14 Live animals for petting and for sale located next to produce displays in a farmers market. Courtesy of Judy Harrison

impossible, to properly clean and sanitize. Both customers and vendors had pets at the market in all three of the trained markets and one of the untrained markets. None of the vendors had handwashing stations at their stands nor were they using gloves or hand sanitizer.

So with all the emphasis on farmers markets and food safety, why are these risky practices still being used? Even as far back as 2004, Yapp and Fairman completed a study of small and medium-sized food businesses in Europe and identified barriers to compliance with food safety regulations [20]. These included time and money and a lack of experience, but more importantly, lack of interest and motivation, lack of trust in food safety legislation and lack of knowledge and understanding. More recently, Parker et al. noted that education alone will not make the food system safer due to barriers such as an underestimation of the effectiveness of prevention strategies and the perceived economic feasibility of implementing risk reduction measures [21]. Some studies suggest more hands-on experiential forms of training may be needed to bring about adoption of safer practices [19].

Before improvements in behavior can be made, however, a bigger question may need to be addressed. How do we get vendors and market managers to take food-borne threats in farmers markets seriously? Until farmers and vendors realize that small, local and organic do not necessarily mean "safe," adoption of better practices may be limited.

However, some market managers and/or individual vendors have adopted best practices for keeping foods safe. These include displaying foods on tables and not on the ground; storing and displaying foods in plastic containers or stainless steel foodservice pans that can be washed and sanitized between uses; lining straw baskets with fabric that can be washed between uses or with paper such as paper towels, freezer paper, deli tissues, etc. that can be discarded after each use to prevent cross-contamination; using tongs or toothpicks for individual sampling; displaying time-temperature control for safety (TCS) foods on ice; providing handwashing and bathroom facilities and posting signage to enforce a "No Pets" policy in the market. In some markets, allergen labeling has been observed [11].

A more detailed discussion of risk analysis in farmers markets is presented in Chap. 9, and a more detailed discussion of model practices for farmers markets is presented in Chap. 10.

Summary

Studies have identified the potential for food safety hazards to exist in farmers markets. Self-reported data as well as direct observations have noted problems with hand hygiene, sanitation, temperature control and animals in the market place. Surveys of regulatory personnel and food safety educators have identified a perceived lack of knowledge among vendors with whom they work about food safety risks associated with products being produced and sold, a lack of understanding about regulatory requirements and licensing that may be needed and a lack of knowledge related to food allergens and allergen labeling. Risky behaviors have been widely observed in markets across the U.S. as well as in other countries. This raises questions and concerns about products that consumers view as more nutritious and safer than those sold in grocery stores.

References

1. Centers for Disease Control and Prevention (2009) Trichinellosis Surveillance – United States, 2002–2007. Surveillance Summaries. MMWR. [cited 2016 Sept 16]; 58 (SS-9). Available from https://www.cdc.gov/mmwr/preview/mmwrhtml/ss5809a1.htm
2. Falkenstein D (2010) Iowa *Salmonella* outbreak linked to Mexican food products. Food Poison Journal. July [cited 2016 Sept 16]. Available from http://www.foodpoisonjournal.com/foodborne-illness-outbreaks/iowa-salmonella-outbreak-linked-to-mexican-food-products/#.V7YWtzUYkSy
3. Goetz G (2011) Did deer cause Oregon's strawberry outbreak. [cited 2016 Sept16]. Available from http://www.foodsafetynews.com/2011/08/epis-pinpoint-strawberries-in-or-e-coli-outbreak/#.V970ja3Mnsa
4. News Desk (2012) WA creamery recalls cheese: potential *Listeria* contamination. [cited 2016 Sept16]. Available from http://www.foodsafetynews.com/2012/06/wa-creamery-recalls-cheese-potential-listeria-contamination/#.V98HY63MnsZ
5. Longenberger AH, Palumbo AJ, Chu AK, Moll ME, Weltman A, Ostroff SM (2013) *Campylobacter jejuni* infections associated with unpasteurized milk – multiple states, 2012. Clin Infect Dis 57(2):263–266

6. News Desk (2013) *Salmonella* outbreak linked to cashew cheese sickens 15. [cited 2016 Sept16]. Available from http://www.foodsafetynews.com/2013/12/salmonella-outbreak-linked-to-cashew-cheese-sickens-15/#.V98GNq3MnsZ
7. Burke P, Needham M, Jackson BR, Bokanyi R, St. Germain E, Englender SJ (2016) Outbreak of foodborne botulism associated with improperly Jarred Pesto – Ohio and California, 2014. MMWR. [cited 2016 Sept16] 65(7);175–177. Available from https://www.cdc.gov/mmwr/volumes/65/wr/mm6507a2.htm
8. Levy DJ, Beck NK, Kossik AL, Patti T, Meschke JS, Calicchia M, Hellberg RS (2015) Microbial safety and quality of fresh herbs from Los Angeles, Orange County and Seattle farmers' markets. J Sci Food Agric 95:2641–2645
9. Simonne A, Swisher M, Saunders-Ferguson K (2006) Food safety practices of vendors at farmers' markets in Florida. Food Prot Trends 26(6):386–392
10. U.S. Dept. of Agriculture, National Agricultural Statistics Service (ACH12-7) (2014) Farmers marketing – direct sales through markets, roadside stands and other means up 8 percent since 2007. 2012 Census of agriculture highlights. [cited 2016 Sept16]. Available from https://www.agcensus.usda.gov/Publications/2012/Online_Resources/Highlights/Farmers_Marketing/Highlights_Farmers_Marketing.pdf
11. Harrison J (2016) Personal observations
12. Condra A (2013) Cottage food laws in the United States. Harvard Food Law and Policy Clinic. [cited 2015 Nov 1]. Available from: http://blogs.law.harvard.edu/foodpolicyinitiative/files/2013/08/FINAL_Cottage-Food-Laws-Report_2013.pdf
13. Harrison JA, Critzer FJ, Harrison MA (2016) Regulatory and food safety knowledge gaps associated with small and very small food businesses as identified by regulators and food safety educators – implications for food safety training. Food Prot Trends 36(6):420–427
14. Worsfold D, Worsfold PM, Griffith CJ (2004) An assessment of food hygiene and safety at farmers' markets. Int J Environ Health Res 14(2):109–119
15. Harrison JA, Gaskin JW, Harrison MA, Cannon JL, Boyer RR, Zehnder GW (2013) Survey of food safety practices on small to medium-sized farms and in farmers markets. J Food Prot 76(11):1989–1993
16. Behnke C, Soobin S, Miller K (2012) Assessing food safety practices in farmers' markets. Food Prot Trends 32(5):232–239
17. McIntyre L, Karden L, Shyng S, Allen K (2014) Survey of observed vendor food-handling practices at farmers' markets in British Columbia, Canada. Food Prot Trends 34(6):397–408
18. Vandeputte EG, Pivarnik LF, Scheinberg J, Machado R, Cutter CN, Lofgren IE (2015) An assessment of food safety handling practices at farmers' markets in Rhode Island using a smartphone application. Food Prot Trends 35(6):428–439
19. Pollard S, Boyer R, Chapman B, di Stefano J, Archibald T, Ponder M, Rideout S (2016) Identification of risky food safety practices at southwest Virginia farmers' markets. Food Prot Trends 36(3):168–175
20. Yapp C, Fairman R (2006) Factors affecting food safety compliance within small and medium-sized enterprises: implications for regulatory and enforcement strategies. Food Control 17:42–51
21. Parker JS, DeNiro J, Ivey ML, Doohan D (2016) Are small and medium scale produce farms inherent food safety risks? J Rural Stud 44:250–260

Chapter 4
Food Safety Considerations for Fruit and Vegetable Vendors

Renee R. Boyer and Stephanie Pollard

Abstract Fresh fruits and vegetables are the primary commodity sold at farmers markets in the U.S. The growing popularity of farmers markets and number of foodborne outbreaks linked to produce highlights the need for farmers markets to have a proactive approach to food safety practices to protect farmers, patrons and local economies. Many produce farmers selling at these markets are exempt from the food safety regulations within the Food Safety Modernization Act (FSMA). The lack of regulation for these outlets can lead to gaps in training and understanding of how to manage food safety risks. There may also be increased opportunities for food to become contaminated and/or pathogen growth in these temporary establishment settings because they are outdoor settings that often lack infrastructure such as water and electricity. This chapter provides an overview of fresh produce food safety risks and the current landscape of research surrounding farmers market produce. Recommendations for growers to reduce the risk of selling fresh produce in farmers markets include close attention to growing conditions, food handling, sanitation and prevention of potential cross-contamination.

Keywords Produce • Fresh fruits and vegetables • Temperature control • Water use • Manure use • Worker hygiene • Sanitation • Foodborne illness • Microbial quality • Foodborne pathogens • Farmers market vendor

Fresh produce makes up 82% of the foods for sale at farmers markets [1]. Approximately 76% of farmers market patrons purchase fruits, and 91% purchase vegetables [2]. There has been a steady increase in the number of documented produce associated foodborne outbreaks since 1987. According to data published by the Center for Science in the Public Interest, there were 667 outbreaks and 23,748 illnesses associated with consumption of produce between 2002 and 2011 [3]. Additionally, 46% of confirmed foodborne illnesses, 38% of hospitalizations and

R.R. Boyer (✉) • S. Pollard
Department of Food Science and Technology, Virginia Tech,
401-A HABB1 (0924), 1230 Washington Street SW, Blacksburg, VA 24061, USA
e-mail: rrboyer@vt.edu

© Springer International Publishing AG 2017
J.A. Harrison (ed.), *Food Safety for Farmers Markets: A Guide to Enhancing Safety of Local Foods*, Food Microbiology and Food Safety,
DOI 10.1007/978-3-319-66689-1_4

Table 4.1 Microorganisms linked to select produce outbreaks (2010-present) and recommendations for their control

Pathogen	Common contamination source	Produce linked to outbreaks	Recommendation
Salmonella enterica	Fecal material, improperly composted manure, contaminated water	Sprouts [8–12], Papayas [13], Cantaloupe [14, 15], Cucumbers [16, 17], Mangos [18]	(1) Ensure manure is composted properly (2) Monitor quality of irrigation water (3) Restrict wild animal access from fields (4) Use good sanitation practices
Shiga-toxin producing *Escherichia coli*	Fecal material, improperly composted manure, contaminated water	Sprouts [19, 20], Ready-to-eat salad [21], Romaine lettuce [22, 23], Spinach and spring mix [24]	
Listeria monocytogenes	Soil, environment	Cantaloupe [25], Salads [26]	(1) Use good sanitation practices
Cyclospora	Contaminated water, food handler	Cilantro [27], Fresh produce [28]	(1) Monitor quality of irrigation water
			(2) Exclude ill workers from handling produce

23% of deaths in the United States that were linked to a specific commodity between 1998 and 2008 were attributed to produce [4]. More illnesses were attributed to leafy greens than any other commodity, with greater than two million illnesses between 1998 and 2008 [4].

Produce related outbreaks have specifically captured public attention in the last decade with several high profile multistate outbreaks. Four outbreaks of *Salmonella enterica* associated with consumption of tomatoes resulted in 459 culture-confirmed cases of salmonellosis in 21 states between 2005 and 2006 [5]. In 2006, an outbreak linked to bagged spinach contaminated with *Escherichia coli* O157:H7 resulted in 205 culture confirmed illnesses and three deaths [6]. More recently, in 2014 and 2015, there were at least six multi-state outbreaks, including three linked to sprouts contaminated with *S. enterica* Enteritidis, *Listeria monocytogenes*, or *E. coli* O121, respectively; two linked to cucumbers contaminated with *S. enterica* serovars; and one linked to cilantro contaminated with *Cyclospora* [7]. These are just a few examples of some of the most publicized outbreaks. Some of the most recent foodborne outbreaks linked to produce are listed in Table 4.1. Generally, these examples are larger in scope, with commodities originating on a commercial farm.

Outbreaks, like the ones listed in the table tend to have a high impact since they are multistate and affect a large number of people. It is important to note however, that produce contamination can occur in similar ways regardless of the scope of the farm or producer. It is likely that a direct market grower selling in a local farmers market would have less impact if illnesses were to occur, but the same risks apply and ways to mitigate those risks should be implemented. One common misconception is that produce, grown on a small-scale farm, is automatically safer because it does not travel long distances or come from a large industrial-sized farm. However, regardless of scale, the same risks apply to how fresh produce is grown and handled after harvest.

Foodborne Illness Outbreaks Linked to Farmers Markets

Consumers of local foods have the perception that locally grown produce is safer, and often argue that it has not been linked to foodborne outbreaks (multiple conversations with extension clientele, 2006–2016). It is important to note that tracking foodborne outbreaks which originate from small retail outlets like farmers markets can be extremely difficult. This is due, in part, to the small number of individuals that might consume a contaminated product. This is one reason which may explain the relatively small number of foodborne outbreaks linked directly to produce purchased from farmers markets. Additionally, outbreaks or illnesses originating from localized areas typically receive very little press, therefore there is limited data describing the outbreak, and often limited traceback to determine the exact cause of illness. Nonetheless, several outbreaks have occurred which are notable and will be discussed briefly.

Outbreak #1: Guacamole

An outbreak of *Salmonella enterica* serovar Newport was linked to guacamole and salsa sold at several Iowa farmers markets in the summer of 2010. The products were produced by a local restaurant and sold alongside tamales at over eight different markets across Iowa City, Cedar Rapids and Dubuque County [29]. Forty-four cases of illness were ultimately linked to these products, and at least three people were hospitalized [30]. The exact source of contamination was not identified, but there were several risky food handling behaviors that may have been contributing factors, such as improper sanitation and cross-contamination. Temperature abuse may have also played a role due to warm daytime temperatures and improper temperature control at the markets.

Outbreak #3: Strawberries

An outbreak of *E. coli* O157:H7 was linked to strawberries sold at roadside stands, farm stands and farmers markets in Northwest Oregon [31]. There were 15 cases of illness, four individuals hospitalized, two suffered kidney failure, and one elderly woman died [32]. Traceback investigation linked the strawberries to a local strawberry farm. Markets purchased the strawberries from the farm and resold them to patrons. Early in the investigation, officials encouraged patrons to avoid eating strawberries from any farmers markets in an attempt to reduce further illnesses. The state public health epidemiologist was able to identify the outbreak strain in strawberries from the farm and then further confirm that the source of contamination was deer feces in the strawberry fields.

Microbial Quality of Produce Sold at Farmers Markets

The microbiological quality of produce ranges significantly based on commodity. Tables 4.2 and 4.3 provide a summary of data collected from a number of studies assessing the microbial quality of produce items samples from local markets in the United States and Canada. The methodology used in these studies is not necessarily comparable but can show a range of products and types of tests that have been completed to date. The quality is often evaluated using a number of different microbiological counts including the total aerobic plate count, coliforms, fecal coliforms, generic *Escherichia coli* and *Enterococcus*. This data can provide a general indicator of the quality and cleanliness of a product, with a possible correlation between poor quality (high microbial counts) and presence of pathogens. Total aerobic plate counts (APC) and coliform counts of produce range between about 4 and 7, and 1 and 4 log CFU/g respectively, depending on the product [33–40]. These counts increase with age and handling of the product [33].

Fruits and vegetables sold in outdoor markets are handled and stored with less control over sanitation, worker hygiene and temperature control when compared to supermarkets. For example, many farmers markets have limited access to electricity, handwashing stations, toilets, trash receptacles and cleaning procedures, which may contribute to poor sanitation and hygiene [1, 41]. Additionally, produce are transported to the market in containers/vessels of varying cleanliness including open and enclosed trucks, cold storage, horse trailers and trunks of cars [1]. These practices can contribute to contamination or enhance growth of microorganisms if they are already present.

Regulatory Barriers Associated with Fresh Produce Food Safety

Farmers are encouraged to follow Good Agricultural Practices (GAPs) in order to minimize the occurrence of contamination of fresh fruit and vegetables [43]. These are a set of recommendations and best practices that improve the safety of produce by reducing the risk of contamination occurring on the farm. Produce farmers that sell to national and regional retail chains are commonly required by their buyers to pass a GAPs-based audit. In general, farmers selling in retail agree

Table 4.2 Summary of microbial counts from produce sampled from a variety of local markets in the U.S. and Canada

Produce type	APC	Coliform count Avg count[a] (# positive/# samples)	Generic *E. coli* Avg count[a, b] (# positive/# samples)	References
Basil	–	2.61 (43/52)	1.79 (14/52)	[35]
		4.02[c]	0.86 (ND)[d]	[36]
Berries	5.71	1.63 (2/11)	ND (6/11)	[34]
Carrots	–	–	1.21[b] (9/206)	[37]
Cilantro	–	2.30 (36/41)	1.71 (10/41)	[35]
		4.16[c]	1.13[a] (ND)	[36]
Cucumbers	5.35	3.50 (27/32)	ND (5/32)	[34]
Green onions	6.62	5.72 (10/13)	ND (4/13)	[34]
			1.43[b] (7/129)	[37]
Herbs	6.68	5.21 (11/14)	3/14	[34]
Lettuce	6.88	5.51 (42/46)	ND (20/45)	[34]
			1.25[b] (23/128)	[37]
			82[e] (39/63)	[39]
Lettuce (red leaf)	6.29	1.60	–	[40]
Lettuce (green leaf)	6.11	2.20	–	[40]
Lettuce (romaine)	6.70	1.90	–	[40
	6.04	3.77	–	[38]
Lettuce (bibb)	6.20	3.53	–	[38]
Okra	–	4.05[c]	0.41(ND)	[36]
Parsley	–	2.42 (33/40)	1.96 (8/40)	[35]
Peppers	6.89	4.51 (21/38)	10/36	[34]
Spinach	–	–	1.54[b] (16/59)	[37]
Squash (bitter)	–	3.94[c]	0.36 (ND)	[36]
Squash (stem/leaves)	–	4.02[c]	0.94 (ND)	[36]
Squash (yellow)	6.40	3.63 (36/39)	ND (9/39)	[34]
Tomatoes	6.93	3.98 (37/64)	ND (18/64)	[34]
	–		0.425[c](4/63)	[39]
Yardlong beans	–	3.82[c]	0.31 (ND)	[36]
Zucchini	6.11	3.00 (14/16)	ND (2/16)	[34]

[a]Expressed as log CFU/g unless otherwise noted
[b]Log MPN/g
[c]Fecal coliforms (log CFU/g)
[d]Not determined
[e]MPN/100 mL

that GAPs reduces risk of contamination, however only 60% of farmers surveyed actually follow GAPs recommendations all or most of the time [44]. The Produce Safety Rule under the Food Safety Modernization Act was published November 27th, 2015. This rule requires a set of science based minimum standards for the growing, harvesting, packing and holding of most produce commodities for human consumption [45]. The standards in the rule are based on the GAPs principles and

Table 4.3 Summary of number of produce samples collected from a variety of local markets in the U.S. and Canada that are positive for each pathogens

Produce type	E. coli O157:H7	Salmonella enterica	L. monocytogenes	Campylobacter	Reference
Basil	0/29	0/29	–	–	[36]
		0/52	–	–	[35]
Berries	0/11	0/11	0/11	–	[34]
Cabbage	–	–	–	0/45	[42]
Carrots	0/206	0/206	–	0/206	[37]
	–	–		0/49	[42]
Celery	–	–	–	0/50	[42]
Cilantro	0/9	2/9	–	–	[36]
	–	0/41	–	–	[35]
Cucumbers	0/32	2/32	0/32	–	[34, 42]
	–	–	–	0/43	
Green onions	0/13	3/13	1/13	–	[34]
	0/129	0/129	–	0/129	[37]
	–	–	–	1/40	[42]
Herbs	0/14	3/14	0/14	–	[34]
Lettuce (leafy greens)	0/45	6/45	3/45	–	[34]
	0/128	0/128	–	0/128	[37]
	–	–	–	2/67	[42]
Okra	0/44	3/44	–	–	[36]
Parsley	–	1/40	–	1/42	[35, 42]
	–	–	–	–	
Peppers	0/36	4/36	0/36	–	[34]
Potatoes	–	–	–	1/63	[42]
Radish (w/ leaves)	–	–	–	2/74	[42]
Squash (stems/ leaves)	0/29	8/29	–	–	[36]
Squash (summer)	0/39	2/39	1/39	–	[34]
Squash (bitter)	0/62	0/62	–	–	[36]
Spinach	0/59	0/59	–	0/59	[37]
				2/60	[42]
Tomatoes	0/64	2/64	1/64	–	[34]
	0/120	0/120	–	0/120	[37]
Yardlong beans	0/69	2/69	–	–	[36]

include requirements specific to: (1) worker training, health and hygiene; (2) agricultural water; (3) biological soil amendments of animal origin; (4) domesticated and wild animals; and (5) equipment, tools, and buildings [45]. Small growers that sell an average annual monetary value of $25,000 or less of produce over the previous three year period are exempt from this rule [45]. Those who sell more than

$25,000 but less than $500,000 in food sales may be eligible for a qualified exemption if their sales are directly to end users (defined as directly to consumers, and/or restaurants and retail establishments within the same state or Indian reservation and not more than 275 miles from the farm) and these direct sales exceed sales to all others. In other words, at least 51% of their sales must be to these end users. If eligible for a qualified exemption from the Produce Safety Rule, farmers would be exempt from having to have Produce Safety Alliance Grower Training. It is expected that most vendors selling produce at farmers markets will fall under this exemption. With the emphasis on education and training to deliver information to reduce foodborne illness risk to those growers who are covered under the Produce Safety Rule, there is concern that those growers that are exempt, selling in the direct market, may not seek training.

In the present landscape, as many as 85% of farmers market produce vendors are unaware of GAPs, or believe that the recommendations are not applicable to them, or specifically to farmers market products [1]. These smaller growers have limited access to affordable training, or find the information provided to be confusing and not appropriate for farms of their size. With few curricula specific to small farm operations to educate these individuals; these smaller food businesses have less access to food safety education and training. Modifications to existing curricula to explain requirements under the Produce Safety Rule need to be understandable and appropriate for small scale growers.

Risky Behaviors Reported/Observed Among Produce Vendors Selling at Farmers markets

Limited studies have evaluated food safety knowledge and practices among farmers market produce vendors. A survey conducted in 2010 asked small produce farmers selling in the direct market, and farmers market managers about specific food safety practices used on their farms or in their markets [46]. Data was collected via self-reported survey to provide a snapshot into some of the practices that could be improved with adequate training. Outcomes of this survey are presented in Chaps. 2 and 3.

Since that time, a number of observational studies have been conducted in markets. Using a secret shopper model, where the farmers market vendors do not know that they are being observed, practices across Virginia and North Carolina have been examined. In one study, 42 produce vendors were observed selling produce at farmers markets during the summer of 2014 [34]. Temperature control was identified as a key behavior needing more attention. Fifty-two percent of vendors observed sold pre-cut produce, but only 32% of those used any form of temperature control [34]. Methods of temperature control used included bags of ice, and immersion of products in a pool of iced water, but none used any temperature monitoring methods [34]. Additionally, three vendors offered samples requiring temperature control but provided no cooling method [34]. Potentially hazardous foods like cut melon and cheese, as well as samples, were held without temperature control, and dirty equipment such as knives and other utensils were used to serve food [1]. Other observed behaviors in Virginia included keeping food off of the ground (76%), and using tables with easy to clean surfaces (62%), and using produce bins that were easy to clean (55%) [34]. Similar trends were seen in North Carolina markets.

Recommendations for Fresh Fruit and Vegetable Growers (Pre- and Post-Harvest on the Farm)

The "Guide to Minimize the Microbial Food Safety Hazards for Fresh Fruits and Vegetables" was published in 1998 in response to President Clinton's Initiative to "Ensure the Safety of Imported and Domestic Fruits and Vegetables" [47]. This document outlines Good Agricultural Practices (GAPs) that can be used to reduce the risk of fresh produce being contaminated by foodborne pathogens. These recommendations were a catalyst for extensive research in the fresh produce area, and are the basis for the requirements for the newly published Produce Safety Rule [45], as well as the GAP curriculum developed by Cornell University [48]. Most of the recommendations made to small growers in this section originate from recommendation made in that document [47]. Regardless of size of produce farm, the same risks apply and should be controlled to ensure the safest produce possible. Following are common pre-harvest recommendations for produce growers.

Water Use

Water is essential in farming, but can be a key vehicle for contaminating fresh produce with foodborne pathogens. Ensuring use of high quality potable water sources is essential for reducing risk when growing, processing and harvesting fresh produce. Most farms do not have access to municipal water sources, therefore they use irrigation water that comes from an outside source, such as a well or a pond. Using a well is the most desirable option when municipal water cannot be used because it is an enclosed source with limited opportunity for contamination. The use of surface water (pond, lake or stream) is generally discouraged because the water is open in the environment and can come in contact with sources of contamination such as wildlife. Water application method may also have an influence in transferring pathogens to the plant. Applying water using drip tape (at the base of the roots) is considered an ideal irrigation method because it keeps the water from touching the edible portions of the crop (for those grown above ground). This would be especially important if untreated surface water is used for irrigation.

Regardless of water source, it is essential that the water is tested to ensure its microbial quality. Ideally, the water is free of *E. coli* bacteria. For farms falling under the Produce Safety Rule, surface and well water can be used, but there are rigorous testing requirements, including the establishment of a water quality profile (WQP) with annual sampling after the WQP is established [45]. For direct market-ers that do not fall under this requirement, it is still essential to understand the qual-ity of water used. If small growers who are exempt cannot follow the standards for testing established in the Produce Safety Rule, they should at least test surface water on a quarterly basis and well water annually. According to the Produce Safety Rule, at the time of this publication, water used for overhead and drip irrigation should contain a geometric mean (GM) of 126 CFU or less of generic *E. coli* per 100 mL or a standard threshold value (STV) of 410 or less CFU of generic *E. coli* per 100 mL [45]. Also, at the time of this publication, FDA requires that EPA Method 1603 be used to obtain water quality measurements for growers subject to the Produce Safety Rule. There are currently few labs that offer this test, samples must be kept refrigerated and must reach labs and be analyzed within eight hours. Growers indicate that considerable time and expense are involved with the water testing requirement. There is much discussion and controversy over these requirements and future changes are anticipated. For additional information and to stay abreast of updates on the latest agricultural water requirements, check the FDA website at www.fda.gov [49].

Additional precautions should be taken when rinsing produce post-harvest prior to taking it to market. For the same reasons contaminated water should not be applied to a crop pre-harvest, it is even more important post-harvest. Water sources used for application directly to the produce should be municipal or tested well water suitable for drinking to reduce potential risk to the lowest level possible. Water for applying sprays, washing produce or washing surfaces and equipment post-harvest must contain no detectable levels of *E. coli* [45].

Manure Use

Raw manure of any kind should never be applied directly to edible crops because foodborne pathogens are known to be shed in manure. Raw manure, however, can be applied to fields. At the time of this publication, the Produce Safety Rule does not require specific waiting periods between raw manure application and harvest of crops. However, additional research is being evaluated. Until further guidance becomes available, the FDA has stated that it has no objection to using the National Organic Program recommendations for application of raw manure [45]. The general recommendation in the National Organic Program is to wait 90 days following application of raw manure on fields to harvest crops that are grown above ground (e.g., tomatoes) and to wait at least 120 days to harvest crops that are grown on or in the ground (e.g., cucumbers) [50]. The best practice, would be eliminate this risk by not applying raw manure at all. It is advisable to periodically check the FDA website at www.fda.gov for the latest recommendations regarding use of raw manure. If the manure is properly composted to eliminate pathogens, no waiting period is required.

Properly composted manure can be applied because following proper composting procedures kills pathogens that could be present. For composting to be done correctly, the compost pile should reach temperatures between 131 and 170 °F (55 and 76.7 °C) for a specified time followed by appropriate aging [45, 50]. Depending on the size of the compost pile and method, the pile will need to maintain these temperatures between 3 and 15 days. The minimum size to generate enough heat to properly compost a pile of manure is $5 \times 5 \times 5$ feet ($1.5 \times 1.5 \times 1.5$ m). Composting methods used should be those scientifically proven to eliminate pathogens. Additionally, care should be taken to prevent any cross-contamination between an un-composted pile of manure and properly composted manure, and equipment, tools, etc. that might come in contact with the crop.

Animal Exclusion

The Produce Safety Rule specifies that growers take precautions to avoid harvesting produce that is likely to be contaminated. This means that growers may need to avoid harvesting in areas where animal feces is present. Although the rule does not require animal exclusion, the primary reason to recommend keeping animals out of growing areas for fresh produce on the farm is to reduce the risk of contamination of edible ready-to-eat (RTE) crops with animal feces. Animals can contaminate produce through cross-contamination as well as from defecating in the fields in the vicinity of the crop. Foodborne outbreaks have been traced back to animal contact in the field [32]. Although, extremely difficult, animals can be discouraged from contacting the produce fields through the use of fencing and other less expensive techniques like flashing lights, scarecrows, bird netting and shiny objects to discourage animals from entering the field. Additionally, it is essential to monitor the fields for animal presence and visible signs of potential contamination.

Worker Hygiene

In all aspects of food production, food handlers are considered a significant factor associated with food contamination. Workers on a farm harvesting product in the field are the primary food handlers in the pre-harvest setting. Key control measures in farm and market settings include preventing sick workers from handling food since they have the potential to contaminate it. Regardless of the size of the farming operation, all growers should have a policy that limits contact of sick employees with food.

Adequate handwashing and toilet facilities should also be provided to encourage good hygiene. If the crops are close to the farmers residence, the handwashing and toilet facilities can be within the residence, however, if this is not the case, farmers

Fig. 4.1 Trailer with portable toilet and handwashing facilities to provide easy access for farm workers. Courtesy of Mark Harrison

need to provide handwashing or bathroom facilities, even if they are portable. For handwashing, it can be as simple as supplying a large water container with a spigot, placed on a table with a bucket for the rinse water to go into, soap in a pump dispenser, single use paper towels and a trash can. The toilet facility could be a rented portable toilet that can be moved around the field as worker locations change. Portable units are available that have both toilet and handwashing facilities (Fig. 4.1).

Depending on local health regulations, often handwashing stations or facilities v not required at individual farmers market booths. It is, however, considered best practice to have access to handwashing and to employ frequent handwashing throughout the day. Some markets and health codes will require handwashing facilities in the event that the vendor is cutting (processing) fruits and vegetables at the market. The same measures can be taken to create a handwashing station at the market as previously described for on-farm settings.

Sanitation on the Farm, During Transportation and at the Market

Following good cleaning and sanitation practices is essential for preventing contamination and cross-contamination in all settings. All items that come in contact with the fresh produce directly (buckets, harvesting bins, utensils, tools, equipment, tables at the market etc.) should be made of cleanable materials. A good example of a cleanable surface is a non-porous plastic or metal surface. In food processing, stainless steel is the primary food contact surface, but this may not be practical for farms or farmers markets. An example of surfaces that are not cleanable include wooden surfaces, cardboard boxes and straw baskets. These surfaces tend to hold moisture and are full of cracks and crevices that are hard to clean.

Cleaning and sanitizing are two separate processes. Cleaning is the removal of dirt from a surface, and sanitizing is the reduction in number of disease-causing microorganisms from surfaces. Using both processes is essential in reducing the risk of cross-contamination throughout a location whether it is a packing house or a farmers market stall [51, 52]. There are many different sanitizers that can be used, but the most common and least expensive one is chlorine. Chlorine is typically used at levels of 200 ppm for food contact surfaces, and higher levels for items that do not come in contact with food, such as floors and walls (400 ppm) [51]. If bleach is used as the chlorine source, check the label to determine the percent of hypochlorite. If the concentration is 5.25–6% hypochlorite, mix one tablespoon (15 mL) of the bleach in one gallon (3.8 L) of water to make a 200 ppm solution. If the concentration of the bleach is 8.25%, then use two teaspoons (10 mL) per one gallon (3.8 L) of water to make a 200 ppm solution. If liquid household bleach is used for sanitizing surfaces, it should be unscented bleach with no thickening agents. Chlorine chemistry is complicated. Its efficacy is greatly affected by pH, presence of organic matter, temperature and exposure to air and light [51] In order to ensure that the chlorine sanitizer is effective, solutions should be made fresh every 24 h (or less), be maintained at a pH between 5.0 and 8.0 [51], and the activity should be monitored. This can be done by using chlorine test strips that will measure the free available chlorine using a simple color indicator. Chlorine solutions can be used as a wash, dip or spray for equipment and surfaces, but it is essential to make sure that the sanitizer stays in contact with the surface for at least one minute and is allowed to air dry. This ensures that the sanitizer has enough time to be effective. Other sanitizers can be used in fresh produce production and for large-scale production, some are even used in produce wash water. It is essential to make sure that they are food grade, meaning that they are acceptable for use on surfaces that come in contact with food and even the food itself, if they are to be used for flumes, dump tanks, etc. Other sanitizers commonly used are peroxyacetic (or peracetic) acid, quaternary ammonium compounds, and hydrogen peroxide [50, 51].

Temperature Control

Following harvest, proper temperature control is one of the most important practices to prevent foodborne illness. There are some foods that require time/temperature control to ensure their safety (TCS foods) [52]. Most fresh fruits and vegetables are not considered TCS foods, however there are some exceptions. Fresh fruits and vegetables that are considered TCS food include "raw seed sprouts, cut melons, cut leafy greens and cut tomatoes or mixtures of tomatoes that are not modified in any way so that they are unable to support pathogenic microorganism grow or toxin

Table 4.4 Range of practices from riskiest to best[a] used by produce growers on farms (pre- and post-harvest) selling produce at farmers markets

Practice identified	Best practices[a]		Riskiest practice
Water use (irrigation)	Use municipal water or tested well water to irrigate or wash produce	Use untested well water	Use pond surface water
Water use (post-harvest produce washing)	Use municipal water or tested well water to wash produce		Use unpotable, pond water to rinse and wash post-harvest produce
Manure use	Use properly composted manure	Use raw manure with adherence to 90 and 120 day waiting period	Use raw manure
Worker hygiene	Provide handwashing stations and toilet facilities		Provided no handwashing locations and no toilet facilities
Animal exclusion	Takes all measures to exclude animals from produce growing areas • Fencing • Monitoring plan		Uses no form of animal exclusion
Sanitation on farm	Use sanitizer to sanitize harvesting buckets/totes (that are cleanable, plastic surface) or use one-time use new cardboard boxes	Use cleanable harvesting totes/buckets that are infrequently sanitized	Use no sanitation principles

[a]Best practices are those recommended to enhance the safety of products sold

formation" [52]. It is well documented that sprouts, cut melons, cut leafy greens and cut tomatoes support the growth of microorganisms, so holding these items below 41 °F (5 °C) is essential. This includes all points along the way from processing to consumption. During transportation, or at the market these specific foods should be held in a manner that keeps them below 41 °F (5 °C), either using refrigeration, or using an insulated cooler with ice, both of which are monitored with a thermometer to ensure compliance. At the market, the most common ways that these items are distributed are either as samples (for example cut melon samples being offered to patrons to taste) or as cut salad mixes in plastic bags [34]. While other produce types do not have a temperature requirement, it is still a good practice to hold these items at a cool temperature because this reduces microbial growth and preserves quality. Some items, like tomatoes and bananas (to name a few) can suffer from chill injury if kept too cold, so there can be a delicate balance between safety and quality.

Tables 4.4 and 4.5 illustrate a range of produce handling practices identified on farms, during transport to farmers markets and while in the markets ranging from riskiest practices to best practices to enhance product safety.

Table 4.5 Range of practices from riskiest to best[a] used by produce growers during transportation to and at farmers markets

Practice identified	Best practices[a]		Riskiest practice
Sanitation during transport	Sanitize bed of truck or other vessel on a regular basis	Clean bed of truck, but don't use sanitizer	Never clean or sanitize bed of truck
Sanitation at the market	Use a sanitizer to sanitize display tables and containers		Use no sanitation principles
	Use display tables and containers that are cleanable (plastic, metal)		
Handling intact produce at the market	Keep produce as cool as possible without inducing chill injury (which some produce items can get)	Hold produce at ambient temperature	Hold at ambient temperature
	Display in cleanable containers (non-porous plastic)	Display in clean able containers (non-porous plastic)	Display in cardboards or wooden boxes on the ground
Handling cut produce at the market	Used refrigeration or a cooler with potable ice that is monitored with a thermometer	Used refrigeration or a cooler with potable ice	No cooling method used
Handling samples at the market	Prepares samples at home in a sealable container, transports to market and holds on ice during sampling	Prepare samples at the market using best practices (incorporating handwashing station, portable three compartment sink etc…). Using temperature control	Use no temperature control or sanitation practices

[a]Best practices are those recommended to enhance the safety of products sold

Summary

Studies, both survey and observational, have identified a range of practices from risky to best practices to enhance product safety being used by farmers selling fresh produce at farmers markets. Outbreaks of foodborne illnesses associated with fresh produce have led to a nationwide standard for growing, harvesting and handling of produce under the Produce Safety Rule that is a part of the Food Safety Modernization Act. However, many vendors at farmers markets may be exempt from this rule. Even if exempt, anyone selling fresh produce for human consumption should be using good agricultural practices to avoid contamination of their products with foodborne pathogens and should transport and hold products at the market using best practices to prevent foodborne illnesses. These practices include proper handling of manure and/or compost; monitoring water quality of both pre-harvest and post-harvest water; ensuring good health and hygiene practices of workers handling produce; proper sanitation of facilities and equipment; and proper storage, transportation and holding temperatures of products to minimize bacterial growth.

References

1. Smathers SA, Phister T, Gunter C, Jaykus L, Oblinger J, Chapman B (2012) Evaluation, development and implementation of an education curriculum to enhance food safety practices at North Carolina farmers markets. Master's Thesis, North Carolina State University [cited 2016 Mar 16]. Available from: http://repository.lib.ncsu.edu/ir/bitstream/1840.16/8094/1/etd.pdf
2. Andreatta S, Wickliffe W (2002) Managing farmer and consumer expectations: a study of a North Carolina farmers market. Hum Organ 61(2):167–176
3. Center for Science and the Public Interest [Internet]. Washington DC (2014) Outbreak alert! 2014: a review of foodborne illness in America from 2002–2011 [cited 2016 Mar 15]. Available from: http://cspinet.org/reports/outbreakalert2014.pdf
4. Painter JA, Hoekstra RM, Ayers R, Tauxe RV, Braden CR, Angulo FJ et al (2013) Attribution of foodborne illnesses, hospitalizations, and deaths to food commodities by using outbreak data, United States, 1998-2008. Emerg Infect Dis 19(3):407–415
5. Centers for Disease Control and Prevention (2007) Multistate outbreaks of *Salmonella* infections associated with raw tomatoes eaten in restaurants – United States, 2005—2006 56(3):909–911
6. Grant J, Wendelboe AM, Wendel A, Jepson B, Torres P, Smelser C et al (2008) Spinach-associated *Escherichia coli* O157:H7 outbreak, Utah and New Mexico, 2006. Emerg Infect Dis 14(10):1633–1636
7. Centers for Disease Control and Prevention [Internet] Atlanta GA (2016) List of selected multistate foodborne outbreak investigations [cited 2016 Mar 15]. Available from: http://www.cdc.gov/foodsafety/outbreaks/multistate-outbreaks/outbreaks-list.html
8. Centers for Disease Control and Prevention [Internet] (2010) Multistate outbreak of human *Salmonella* Newport infections linked to raw alfalfa sprouts [cited 2016 Mar 15]. Available from: http://www.cdc.gov/salmonella/2010/newport-alfalfa-sprout-6-29-10.html
9. Centers for Disease Control and Prevention [Internet]. Atlanta, GA (2011) Multistate outbreak of Human *Salmonella* I 4,[5],12:i:- infections linked to alfalfa sprouts [cited 2016 Mar 15]. Available from: http://www.cdc.gov/salmonella/2010/alfalfa-sprouts-2-10-11.html
10. Centers for Disease Control and Prevention (2011) Multistate outbreak of human *Salmonella* Enteritidis infection linked to alfalfa sprouts and spicy sprouts [cited 2016 Mar 15]. Available from: http://www.cdc.gov/salmonella/2011/alfalfa-spicy-sprouts-7-6-2011.html
11. Centers for Disease Control and Prevention (2015) Multistate outbreak of *Salmonella* Enteritidis infections linked to bean sprouts [cited 2016 Mar 15]. Available from: http://www.cdc.gov/salmonella/enteritidis-11-14/index.html
12. Centers for Disease Control and Prevention (2016) Multistate outbreak of *Salmonella* Muenchen infections linked to alfalfa sprouts produced by Sweetwater farms [cited 2016 Mar 16]. Available from: http://www.cdc.gov/salmonella/muenchen-02-16/index.html
13. Centers for Disease Control and Prevention (2011). Multistate outbreak of human *Salmonella* Agona infections linked to whole, fresh imported papayas [cited 2016 Mar 15]. Available from: http://www.cdc.gov/salmonella/2011/papayas-8-29-2011.html
14. Centers for Disease Control and Prevention (2011) Multistate outbreaks of *Salmonella* Panama infections linked to cantaloupe [cited 2016 Mar 15]. Available from: http://www.cdc.gov/salmonella/2011/cantaloupes-6-23-2011.html
15. Centers for Disease Control and Prevention (2012) Multistate outbreak of *Salmonella* Typhimurium and *Salmonella* Newport infections linked to cantaloupe [cited 2016 Mar 15]. Available from: http://www.cdc.gov/salmonella/typhimurium-cantaloupe-08-12/index.html
16. Centers for Disease Control and Prevention (2013) Multistate outbreak of *Salmonella* St. Paul infections linked to imported cucumber [cited 2016 Mar 15]. Available from: http://www.cdc.gov/salmonella/saintpaul-04-13/index.html
17. Angelo KM, Chu A, Anand M, Nugyen TA, Bottichio L, Wise M et al (2015) Outbreak of *Salmonella* Newport Infections linked to cucumbers – United States, 2014. MMWR 64(06):144–147
18. Centers for Disease Control and Prevention (2012) Multistate outbreak of *Salmonella* Braenderup infections associated with mangoes [cited 2016 Mar 15]. Available from: http://www.cdc.gov/salmonella/braenderup-08-12/index.html

19. Centers for Disease Control and Prevention (2014) Multistate outbreak of Shiga-toxin producing *Escherichia coli* O121 infections linked to raw clover sprouts [cited 2016 Mar 16]. Available from: http://www.cdc.gov/ecoli/2014/O121-05-14/index.html

20. Centers for Disease Control and Prevention (2016) Multistate outbreak of Shiga-toxin producing *Escherichia coli* O157 infections linked to alfalfa sprouts produced by Jack & The Green Sprouts [cited 2016 Mar 16]. Available from: http://www.cdc.gov/ecoli/2016/o157-02-16/index.html

21. Centers for Disease Control and Prevention (2013) Multistate outbreak of Shiga-toxin producing *Escherichia coli* O157:H7 infections linked to ready to eat salads [cited 2016 Mar, 16]. Available from: http://www.cdc.gov/ecoli/2013/O157H7-11-13/index.html

22. Centers for Disease Control and Prevention (2010) Multistate outbreak of Shiga-toxin producing *Escherichia coli* O145 infections linked to shredded romaine lettuce from a single processing facility [cited 2016 Mar 16]. Available from: http://www.cdc.gov/ecoli/2010/shredded-romaine-5-21-10.html

23. Centers for Disease Control and Prevention (2012) Multistate outbreak of Shiga-toxin producing *Escherichia coli* O157:H7 infections linked to romaine lettuce [cited 2016 Mar 16]. Available from: http://www.cdc.gov/ecoli/2011/romaine-lettace-3-23-12.html

24. Centers for Disease Control and Prevention (2012) Multistate outbreak of Shiga-toxin producing *Escherichia coli* O157:H7 infections linked to organic spinach and spring mix blend [cited 2016 Mar 16]. Available from: http://www.cdc.gov/ecoli/2012/O157H7-11-12/index.html

25. Cosgrove S, Cronquist A, Wright G, Ghosh T, Vogy R, Teitell P et al (2011) Multistate outbreak of listeriosis associated with Jensen Farms cantaloupe – United States, August–Sept. 2011. MMWR 60(39):1357–1358

26. Centers for Disease Control and Prevention (2016) Multistate outbreak of listeriosis linked to packaged salads produced at Springfield, Ohio Dole processing facility [cited 2016 Mar 16]. Available from: http://www.cdc.gov/listeria/outbreaks/bagged-salads-01-16/index.html

27. Centers for Disease Control and Prevention (2015) Cyclosporiasis outbreak investigations – United States, 2014 [cited 2016 Mar 16]. Available from: http://www.cdc.gov/parasites/cyclosporiasis/outbreaks/2014/index.html

28. Centers for Disease Control and Prevention (2013) Cyclosporiasis outbreak investigations – United States, 2013 [cited 2016 Mar 16]. Available from: http://www.cdc.gov/parasites/cyclosporiasis/outbreaks/investigation-2013.html

29. Ohlemeier D (2011) Iowa Salmonella outbreak traces to guacamole and salsa. The Packer [Internet] [cited 2016 Mar 16]. Available from: http://www.thepacker.com/fruit-vegetable-news/fresh-produce-retail/iowa_salmonella_outbreak_traced_to_guacamole_and_salsa_122012119.html

30. Iowa Department of Public Health (2010) Iowa surveillance of notifiable and other diseases [cited 2017 Jul 3]. Available from: https://idph.iowa.gov/Portals/1/Files/CADE/IDPH_Annual_Rpt_2010_final.pdf

31. Terry L (2011) Oregon confirms deer droppings caused *E. coli* outbreak tied to strawberries. The Oregonian/OregonLive. August 17 [cited 2017 Jul 3] The Oregonian/OregonLive. Available from: http://www.oregonlive.com/washingtoncounty/index.ssf/2011/08/oregon_confirms_deer_droppings.html

32. Laidler MR, Tourdjman M, Buser GL, Hostetler T, Repp KK, Leman R, Samadpour M, Keene WE (2013) *Escherichia coli* O157:H7 infections associated with consumption of locally grown strawberries contaminated by deer. CID 57:1129–1134

33. Johnston LM, Jaykus L, Moll D, Martinez MC, Ancisso J, Mora B, Moe CL (2005) A field study of the microbiological quality of fresh produce. J Food Prot 68(9):1840–1847

34. Pollard, S., R. Boyer, B. Chapman, J. di Stefano, T. Archibald, M. Ponder, S. Rideout (2016) Identification of risky food safety practices at Virginia's farmers markets and development of a food safety plan to help farmers market managers. Virginia Tech Electronic thesis and dissertation. [cited 2016 March 16]. Available from: https://vtechworks.lib.vt.edu/handle/10919/78196

35. Levy DJ, Beck NK, Kossik AL, Patti T, Meschke JS, Calicchia M et al (2015) Microbial safety and quality of fresh herbs from Los Angeles, Orange County and Seattle farmers' markets. J Sci Food Agric 95(13):2641–2645

36. Pan F, Li X, Carabez J, Ragosta G, Fernandez KL, Wang E et al (2015) Cross-sectional survey of indicator and pathogenic bacteria on vegetables sold from Asian vendors at farmers' markets in Northern California. J Food Prot 78(3):602–608

37. Bohaychuk VM, Bradbury RW, Dimock R, Fehr M, Gensler GE, King RK et al (2009) A microbiological survey of selected Alberta-grown fresh produce from farmers' markets in Alberta Canada. J Food Prot 72(2):415–420

38. Soendjojo E (2012) Is local produce safer? Microbiological quality of fresh lettuce and spinach from grocery stores and farmers markets. J Purdue Undergr Res [cited 2015 Mar 17] 2:54–63. Available from: http://docs.lib.purdue.edu/cgi/viewcontent.cgi?article=1022&context=jpur. Accessed 8 Nov 2012

39. Leang A, Meschke JS, Cangelosi GA, Beck NK (2013) Prevalence of *Salmonella* and *E. coli* on produce from Seattle farmers markets. Masters Thesis, University of Washington, Seattle, WA [cited 2016 Mar 16]. Available from: https://digital.lib.washington.edu/researchworks/bitstream/handle/1773/23403/Leang_washington_0250O_11796.pdf?sequence=1&isAllowed=y

40. Wood J, Chen J, Friesen E, Delaquis P, Allen K (2015) Microbiological survey of locally grown lettuce sold at farmers' markets in Vancouver, British Columbia. J Food Prot 78(1):203–208

41. Worsfold D, Worsfold PM, Griffith CJ (2004) An assessment of food hygiene and safety at farmers' markets. Int J Environ Health Res 14(2):109–119

42. Park CE, Sanders GW (1992) Occurrence of thermotolerant *Campylobacter* in fresh vegetables sold at farmers' outdoor markets and supermarkets. Can J Microbiol 38(4):313–316

43. Parker J, Wilson R, Lejeune JT, Rivers L, Doohan D (2012) An expert guide to understanding grower decisions related to fresh fruit and vegetable contamination prevention and control. Food Control 26:107–116

44. Ivey ML, Lejeune JT, Miller SA (2012) Vegetable producers' perceptions of food safety hazards in the Midwestern USA. Food Control 26:453–465

45. U.S. Food and Drug Administration (2016) FSMA final rule on produce safety [Internet] Washington, DC [cited 2016 Mar 16]. Available from: http://www.fda.gov/Food/GuidanceRegulation/FSMA/ucm334114.htm

46. Harrison JA, Gaskin JW, Harrison MA, Cannon JL, Boyer RR, Zehnder GW (2013) Survey of food safety practices on small to medium-sized farms and in farmers' markets. J Food Prot 76(11):1989–1993

47. U.S. Food and Drug Administration (1998) Guide to minimize microbial food safety hazards for fresh fruits and vegetables [Internet] [cited 2016 Mar 16]. Available from: http://www.fda.gov/downloads/Food/GuidanceRegulation/UCM169112.pdf

48. Cornell University (2016) National Good Agricultural Practices Program [Internet]. Ithaca, NY [cited 2016 Mar 17]. Available from: http://www.gaps.cornell.edu/

49. U.S. Food and Drug Administration (2017) Equivalent testing methodologies for agricultural water [cited 2017 Sept 21]. Available from: https://www.fda.gov/Food/FoodScienceResearch/LaboratoryMethods/ucm575251.htm

50. Code of Federal Regulations Title 7, subtitle B, Chapter I, Subchapter M, part 205 (2016) [cited 2016 Mar 17]. Available from: http://www.ecfr.gov/cgi-bin/text-idx?SID=50858264a6a920c0c46496781ccee930&mc=true&node=pt7.3.205&rgn=div5

51. Beuchat LR (2000) Use of sanitizers in raw fruit and vegetable processing. In: Alzamora SM, Tapia MS, Lopez-Malo A (eds) Minimally processed fruits and vegetables. Aspen Publishers, Gaithersburg, MD, pp 63–78

52. U.S. Food and Drug Administration (2013) Food Code [Internet]. College Park, MD [cited 2016 Mar 17]. Available from: http://www.fda.gov/downloads/Food/GuidanceRegulation/RetailFoodProtection/FoodCode/UCM374510.pdf

Chapter 5
Food Safety Considerations for Meat and Poultry Vendors

Faith J. Critzer

Abstract Fresh meat and poultry have been long recognized as sources of food-borne pathogens. These products have become increasingly present at farmers markets as customers seek to fulfill their entire shopping list with locally sourced foods and vendors look to diversify their offerings. With this growing segment of the market comes a need to understand what food safety risks may be associated with these products when distributed through farmers markets. The regulatory landscape in the U.S. is covered in this chapter to provide a better understanding of the food safety-based regulations locally sourced meat and poultry producers must follow. Exemptions of small-scale local producers for certain products also will be are addressed. Best practices which should be considered by vendors of meat and poultry products are presented.

Keywords Federal Meat Inspection Act • Poultry Products Inspection Act • USDA • *Salmonella* • *Campylobacter* • HACCP • Temperature control

While the farmers market is dominated by vendors offering sale of produce, many vendors have begun to fill gaps where competition is not as prevalent. One of these areas is the sale of meat and poultry. The 2014 USDA National Farmers Market Manager Survey indicated that 62% of market managers were actively seeking vendors who were offering products other than fruits and vegetables [1]. This chapter examines some of the regulatory oversight in the U.S. involved with processing these products, evidence for potential food safety risks, and common considerations that should be given to those wishing to safely offer meat and poultry for sale at farmers markets.

F.J. Critzer (✉)
Department of Food Technology and Science, University of Tennessee,
103 Food Safety and Processing Building, 2600 River Drive, Knoxville, TN 37996, USA
e-mail: faithc@utk.edu

© Springer International Publishing AG 2017   57
J.A. Harrison (ed.), *Food Safety for Farmers Markets: A Guide to Enhancing Safety of Local Foods*, Food Microbiology and Food Safety,
DOI 10.1007/978-3-319-66689-1_5

Regulatory Considerations

The Federal Meat Inspection Act (FMIA; 21 U.S.C. 601, et seq.) and Poultry
Products Inspection Act (PPIA; 21 U.S.C. 451, et seq.) provide a national frame-
work for the safe processing of meat and poultry. Both acts have provisions that
allow states operating their own inspection programs equivalent to the FMIA and
PPIA with respect to (1) antemortem and postmortem inspection, (2) re-inspection,
(3) sanitation and (4) recordkeeping provisions to operate under a state cooperative
inspection program. Meat processed under these cooperative agreements are limited
to intra-state commerce. Currently, 27 states operate their own meat and poultry
inspection programs as shown in Fig. 5.1 [2].

Talmadge-Aiken plants are another type of cooperative agreement between states
and the USDA. The Talmadge-Aiken act of 1962 (7 U.S.C. 450) allows for states to
conduct federal inspections on meat and poultry through cooperative agreements
with the USDA. The products that pass inspection bear the federal mark of inspection, such as that shown for a poultry product in Fig. 5.2. This inspection agreement
was established to allow for federal inspection to occur in remote locations where it
would not be cost effective to staff federal inspectors. Currently, Alabama, Georgia,
Illinois, Mississippi, North Carolina, Oklahoma, Texas, Utah and Virginia coopera-
tively operate under the Talmadge-Aiken Act.

While meat (beef, pork, sheep) must be processed in either a state or federally
inspected facility, there is an exemption for poultry processing inspections at the
federal level. When reviewing these exemption scenarios for poultry, it is important
to consider if a state has laws prohibiting the sale of poultry that is not inspected.
There are four exemptions detailed for poultry in the Poultry Products Inspection
Act (PPIA; 9 C.F.R 381.10) and included in the USDA-FSIS *Guidance for*

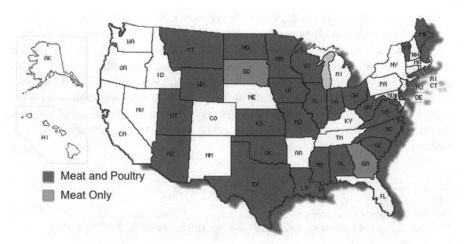

Fig. 5.1 States that currently have state cooperative meat and poultry inspection programs (http://
www.fsis.usda.gov/OPPDE/rdad/FSISDirectives/5720-2Rev3.pdf)

Fig. 5.2 Federal mark of
inspection for a poultry
product. All meat and
poultry which has been
inspected by the USDA-
FSIS or a state cooperative
agreement operating under
the Talmadge-Aiken Act
will bear a federal mark of
inspection (Photo: USDA).
Photo obtained from
Flickr- creative commons
license photo for: https://
www.flickr.com/photos/
usdagov/6691150863/in/
photolist-7kVCQY-
bcgU34-bcgUvc-bcgTAc-
ee8Zou-ewUFCx-ee3ioZ-
ewXRnj-oANNQF-
nsKxEW-oANLhK-
ewXQBU-oRgSnu-
oRgSXC-e2fP7Z-h5NLeT-
oANMyx-h5Q5WD-
h5P61Q-rjCPyJ-h5NThw-
e2fR5r-dabucK-daoex6-
CEkW2f-AKRmBD-
tbvpGE-wXttKD

*Determining Whether a Poultry Slaughter or Processing Operation is Exempt from
Inspection Requirements of the Poultry Products Inspection Act* [3]. These are summarized in Table 5.1.

Prevalence of Foodborne Pathogens Associated with Raw Meat and Poultry Sold at Farmers Markets

Many poultry manufacturers are taking advantage of exemptions described in
Table 5.1 as a means to sell their product direct to consumers at farmers markets.
Researchers are just beginning to determine the contamination rates of raw poultry
processed under this exemption, given that slaughter, processing, packaging and
distribution are done without the oversight of USDA inspection.

A study by Scheinberg et al. evaluated the occurrence of *Salmonella* and
Campylobacter spp. on raw poultry processed under the exemption and sold at
Pennsylvania area farmers markets as compared to organic and conventional poultry

Table 5.1 Federal poultry processing exemptions pertinent for farmers market commerce (adapted from USDA FSIS Guidebook [3])

Criteria	Size of operation			
	Producer/Grower 1000 bird limit	Producer/Grower 20,000 bird limit	Producer/Grower or other person 20,000 bird limit	Small enterprise
Slaughter limit	1000 per calendar year	20,000 per calendar year		
Further processing	Yes	Yes	Yes	Cut up only
Sell to home consumer	Yes	Yes	Yes	Yes
Sell to any hotel, restaurant and institution	Yes	Yes	Not to institutions	Yes
Sell to distributor	Yes	Yes	No	Yes
Sell to retail store	Yes	Yes	No	Yes
Sale across state lines (Interstate commerce)	No	No	No	No

sold at retail in Pennsylvania [4]. *Campylobacter* and *Salmonella* were more prevalent on poultry processed under the exemption, 90% and 28%, respectively, than conventional poultry sold at retail, 52% and 8%, respectively. No significant differences were observed for *Salmonella* prevalence rates in poultry sold at farmers markets and organically raised and processed poultry, but differences did exist for *Campylobacter*. Another study conducted in Maryland and the District of Columbia found similar *Salmonella* prevalence trends in poultry sampled at farmers markets versus at retail [5].

While some focus has been on prevalence of foodborne pathogen occurrence on poultry sold at farmers markets, there currently are no studies to support any risks linked to meat processed under USDA inspection. Further research should focus on potential food safety risks for meat as well as poultry sold direct to consumers through farmers markets. This information will help drive targeted training on evidence-based risks and help vendors and consumers make educated decisions.

Consumer and Vendor Perceptions of Food Safety Risks Associated with Meat and Poultry Sold at Farmers Markets

Given the exemptions allowed by the PPIA, researchers are beginning to understand what processing methods exempt vendors are utilizing to get their product to market. HACCP-based (Hazard Analysis Critical Control Point) processing methods have been developed based upon sound scientific evidence to decrease the likelihood of foodborne pathogen presence, but processors utilizing the PPIA exemption may have very limited to no knowledge of these practices.

An evaluation of farmers market vendors offering poultry for sale in Pennsylvania markets found that 52% processed their own products [6]. The majority of vendors reported processing in a fixed building, but 38% reported processing outdoors or in a barn. When evaluating practices used for poultry slaughter, 43% reported not using a chemical sanitizer during slaughter. Although 62% reported they chilled poultry to 20–40 °F (approximately −6 to 4 °C), more than a third of all respondents reported that they either did not know the temperature their poultry was chilled to prior to packaging or that it was chilled to 40–60 °F (approximately 4–15 °C). This study also determined that a third of processors were relying upon coolers and ice to keep product cold when processed, and more than 70% were relying upon this method for keeping product cold during transport or at the market.

While this study is very interesting, it is limited based upon the population studied and the focus of their questions. In order to gain a more in-depth assessment of food safety risks, future research should focus on determining knowledge and practices that meat and poultry vendors are using in processing, packing, holding, and distributing their products.

While the majority of farmers market managers are actively recruiting vendors selling products other than produce, food safety concerns may be impacting poultry and meat sales from vendors at farmers markets [1]. In an Oregon-based study, 64% of surveyed farmers market shoppers did not purchase meat or poultry at the market [7]. When asked what was driving this decision, 13% cited food safety concerns. Among farmers market patrons who cited food safety concerns, many focused on temperature abuse concerns, specifically the inability to keep product at proper temperatures using only coolers and ice. Additionally, respondents also cited as a deterrent the fact that the product was not on display so that customers could see it.

Consumers are knowledgeable about certain aspects of food safety, especially with respect to the proper storage temperature of meat and poultry. The results of the Oregon-based study infer that if consumers perceive food safety risks, this will be a driver to keep them from purchasing the product. Market managers and vendors should be aware of these perceived risks and make strides to alter their practices that may not be in alignment with industry standards or food safety messages to consumers.

Best Practices for Meat and Poultry Vendors

Based upon studies of meat and poultry farmers market vendor practices and consumers' perceptions, a series of best practices can be identified to help educators, farmers market mangers and meat and poultry vendors [1, 6, 7]. The following practices focus on those at the market that may result in food safety risks and does not include risks that may have been introduced during the harvesting, processing, or holding of these products.

Temperature control at the market is imperative to maintain the safety of raw meat and poultry. There is a clear risk of temperature abuse among vendors selling meat and poultry at the farmers market without means of adequately maintaining temperature. Based upon the observational data from Pennsylvania, the majority of vendors selling poultry at the farmers market were relying upon coolers with ice (57%) or, in some cases, coolers without ice (14%) [6]. Less than a third of all respondents reported using an electrically powered cooler for maintaining temperature during transport and storage at the market.

The ability of coolers to maintain refrigeration (41 °F, 5 °C) or frozen (0 °F, −18 °C) storage conditions is a serious concern. Farmers markets can have durations of four hours of more, and the air temperature inside the cooler becomes warmer due to frequent opening and closing to remove product for sale. Additionally, the ambient temperature can be quite high in summer months with direct exposure to sunlight further challenging temperature maintenance issues. The use of ice can also be problematic if product is immersed in water as the ice melts, possibly cross-contaminating various cuts stored together if the package is compromised.

In order to demonstrate temperature control, it is necessary for meat and poultry vendors to use thermometers in coolers, refrigerators, or freezers. These thermometers should be checked for accuracy and should be monitored to determine if their means of keeping product temperature controlled at the market is adequate. While a survey of thermometer use among meat and poultry vendors at farmers markets has not been conducted, a survey of produce vendors at farmers markets in Virginia revealed that none of the participants used thermometers [8].

Freezers or refrigerators operated by an inverter are better equipped to maintain temperature while at the market (Fig. 5.3). These units can be stored on trucks and used to display cuts of meat that will not enter commerce, but showcase the products vendors are offering for sale. Even when using electrically powered freezers and coolers, thermometers must be used to monitor temperature. If using coolers, dry ice is a much better alternative than ice given that it evaporates and dissipates not posing the same risk of cross-contamination as melting ice.

Safe handling instructions to communicate with consumers is also necessary. The Food Code 2013 from the U.S. Food and Drug Administration (FDA) details the need for appropriate labeling of any meat or poultry as detailed in the Code of Federal Regulations in 9 CFR 317.2(l) and 9 CFR 381.125(b) [9]. An example of the text that should appear in the Safe Handling Instructions Label is shown in Fig. 5.4.

Of course, poultry processed under the PPIA exemptions has not been inspected by the USDA, and therefore must be labeled as follows:

Safe Handling Instructions

Some food products may contain bacteria that could cause illness if the product is mishandled or cooked improperly. For your protection, follow these safe handling instructions.

- Keep refrigerated or frozen. Thaw in refrigerator or microwave.
- Keep raw meat and poultry separate from other foods. Wash working surfaces (including cutting boards), utensils, and hands after touching raw meat or poultry.

Fig. 5.3 Freezers operating from an inverter are used at a farmers market to improve the likelihood product will remain at proper storage temperature (Courtesy of Faith Critzer)

Safe Handling Instructions

This product was prepared from inspected and passed meat and/or poultry. Some food products may contain bacteria that could cause illness if the product is mishandled or cooked improperly. For your protection, follow these safe handling instructions.

Keep refrigerated or frozen. Thaw in refrigerator or microwave.

Keep raw meat and poultry separate from other foods. Wash working surfaces (including cutting boards), utensils, and hands after touching raw meat or poultry.

Cook thoroughly.

Keep hot foods hot. Refrigerate leftovers immediately or discard.

Fig. 5.4 Safe handling instructions that should appear on USDA-inspected meat and poultry (Photo: USDA)

- Cook thoroughly.
- Keep hot foods hot. Refrigerate leftovers immediately or discard.

Assuring the product is not temperature abused during distribution at the farmers market and communicating food safety best practices via safe handling instructions to consumers are two practices that can greatly impact the safety of the meat and poultry offered for sale at farmers markets. Of course, events that occur prior to the market during the harvesting, processing and holding of meat and poultry can impact food safety dramatically. While the oversight of state or federal inspection during these processes does not assure proper application of food safety best practices, having a well-developed hazard analysis critical control point (HACCP) plan, which is mandatory under the FMIA and PPIA, does decrease these risks. Therefore, it is imperative that processors using the PPIA exemption for selling poultry make strides towards educating themselves and developing a HACCP plan for use in their operation to enhance product safety.

Summary

Many vendors wishing to diversify their offerings for sale at the farmers market are already selling meat and poultry products. Safety concerns arise for all products in this group based upon the risk of improper holding temperatures that may allow for growth of foodborne pathogens. Best practices should include monitoring temperature while in transport and at the market to assure that refrigerated or frozen goods are kept at or below 41 or 0 °F (5 or −18 °C), respectively. Actively communicating with customers on proper handling of meat and poultry purchased at the market will also decrease the likelihood of food safety issues arising.

Additional risks arise from allowed sales of poultry products not been processed under USDA or state equivalent inspection programs through a series of exemptions in the PPIA. The lack of developed food safety best practices in these situations has been evaluated in a limited manner and appears to be resulting in a higher prevalence of foodborne pathogens. Additionally, it appears that processors operating under this exemption do not have adequate knowledge of food safety systems that should be in place when processing poultry. Subsequent research should further substantiate these findings and focus on education for these vendors to improve their practices.

References

1. U.S. Dept. of Agriculture, Agricultural Marketing Service (2015) 2014 National Farmers Market Manger Survey Summary [cited 2016 Nov 30]. Available from: http://www.ams.usda. gov/sites/default/files/media/2014%20Farmers%20Market%20Managers%20Survey%20 Summary%20Report%20final%20July%2024%202015.pdf

2. U.S. Dept. of Agriculture, Food Safety and Inspection Service (2015) States Operating their Own MPI Programs [updated March 23, 2015, cited 2016 Nov 30]. Available from: http://www.fsis.usda.gov/wps/portal/fsis/topics/inspection/state-inspection-programs/state-inspection-and-cooperative-agreements/states-operating-their-own-mpi-programs
3. U.S. Dept. of Agriculture, Food Safety and Inspection Service (2006) Guidance for determining whether a poultry slaughter or processing operation is exempt from inspection requirements of the poultry products inspection act [cited 2017 Apr 13]. Available from: https://www.fsis.usda.gov/wps/wcm/connect/0c410cbe-9f0c-4981-86a3-a0e3e3229959/Poultry_Slaughter_Exemption_0406.pdf?MOD=AJPERES
4. Scheinberg J, Doores S, Cutter CN (2013) a microbiological comparison of poultry products obtained from farmers' markets and supermarkets in Pennsylvania. J Food Saf 33(3):259–264
5. Peng M, Salaheen S, Almario JA, Tesfaye B, Buchanan R, Biswas D (2016) Prevalence and antibiotic resistance pattern of *Salmonella* serovars in integrated crop-livestock farms and their products sold in local markets. Environ Microbiol 18(5):1654–1665
6. Scheinberg J, Radhakrishna R, Cutter CN (2013) Food safety knowledge, behavior, and attitudes of vendors of poultry products sold at Pennsylvania farmers' markets. J Ext 51(6):6FEA4
7. Gwin L, Lev L (2011) Meat and poultry buying at farmers' markets: a survey of shoppers at three markets in Oregon. J Ext 49(1):1RIB4
8. Pollard S, Boyer R, Chapman B, di Stefano J, Archibald T, Ponder M, Rideout S (2016) Identification of risky food safety practices at southwest Virginia farmers' markets. Food Prot Trends 36(3):168–175
9. U.S. Food and Drug Administration (2015) Food code 2013 [cited 2017 Apr 13]. Available from: https://www.fda.gov/food/guidanceregulation/retailfoodprotection/foodcode/ucm374275.htm

Chapter 6
Food Safety Considerations for All Other Foods Sold at Farmers Markets

Faith J. Critzer

Abstract Value-added foods are a multibillion-dollar industry, where a vast array of products are being sold at farmers markets across the United States. The ingenuity of vendors has truly expanded the number of processed foods sold at farmers markets, but there are some staples that are more commonly found. This chapter's focus has been placed on those items most commonly found at farmers markets such as jarred shelf-stable foods, breads, cookies, eggs, juice and cider. For each product category, a discussion of food safety concerns, intrinsic and processing controls commonly used and their linkage to food safety, pertinent U.S. regulations, and recommended processing guidelines to mitigate risk and control for foodborne pathogen growth are included.

Keywords Processed foods • Cottage foods • Regulations • Foodborne pathogens • Allergen

More than 100,000 farms engaged in selling value-added foods at farmers markets that were valued at more than $3.9 billion, based upon the 2015 USDA Local Food Marketing Practices Survey [1]. There are an array of products sold at farmers markets in raw or ready-to-eat form ranging from preserves to fresh baked bread. With such an assortment of goods sold, it can be hard for vendors and market managers to assess risks and take the appropriate steps to mitigate food safety issues where they may arise. Additionally, as discussed in Chap. 3, there is an increasing trend with the cottage food laws sweeping across the United States to deregulate foods sold direct to consumers locally. With this trend, there will be an even larger burden

F.J. Critzer (✉)
Department of Food Technology and Science, University of Tennessee,
103 Food Safety and Processing Building, 2600 River Drive, Knoxville, TN 37996, USA
e-mail: faithc@utk.edu

© Springer International Publishing AG 2017
J.A. Harrison (ed.), *Food Safety for Farmers Markets: A Guide to Enhancing Safety of Local Foods*, Food Microbiology and Food Safety,
DOI 10.1007/978-3-319-66689-1_6

on manufacturers and market managers to become knowledgeable on food safety risks and seek out information on how to prepare and hold foods.

With several diverse foods occupying this category, the following categories have been used to better discuss food safety risks, regulatory requirements, and approaches to mitigate risks when preparing or holding foods for sale at farmers markets.

- Eggs
- Shelf-stable jarred foods (pickles, salsa, jams, jellies, sauces, marinades)
- Juice and cider
- Baked goods (bread, cookies, crackers, snack mixes)

Eggs

In the U.S., there are federal regulations that govern requirements for farms selling shell eggs. Federal regulation implemented by the FDA is known as the Egg Safety Rule (21 CFR Part 118). This regulation is aimed at safe production of eggs and has requirements to specifically reduce the contamination of shell eggs with *Salmonella* Enteritidis. The second set of federal regulations are based on the Egg Products Inspection Act (EPIA; 21 CFR Part 15) which allows for inspections of farms packing their own eggs to assure U.S. standards for shell eggs are being properly applied to eggs entering commerce. Many farms will find that they are exempt from these federal regulations because they have less than 3000 laying hens. Some states will have laws governing fresh shell egg sales, but even in cases where there are no regulatory requirements, farmers should still have operating practices in place that focus on proper cleaning and storage of eggs. This is because eggs are held in the cloaca prior to laying, and as such, they are easily contaminated with feces. It is well established that shell eggs can be a source of *Salmonella*, with *Salmonella* Enteritidis being one of the most common serovars.

Research has shown that management practices greatly impact *Salmonella* contamination rates on shell eggs. It is also understood that management practices vary greatly between large and small flocks. When evaluating small (<3000 layers) and medium (3000–31,000 layers) sized flocks using cage free systems, seven of the forty participating farms (17.5%) were positive for *Salmonella* Enteritidis from environmental samples [2]. Only small farms were found to have both positive environmental and egg samples. The primary management strategies associated with decreased *Salmonella* positive environmental and egg samples were proper rodent control, as mice were the primary vector for spreading *Salmonella* [2]. Additionally, a *Salmonella* vaccination program within the flock also worked well in managing the risk of contamination for cage free management systems in both small and medium farms.

Shelf-Stable Jarred Foods

Many manufactured foods prepared for sale at farmers markets fit within this category. It is an attractive option for value-added products to be made by farmers who have excess or slightly blemished fruits and vegetables and would like to divert these goods from potential waste if not sold in the fresh market. Additionally, these foods have a long shelf life and can provide an array of goods at the stand during the beginning and end of the season when the fresh crops are not in full abundance. The sauce and marinade segment has always been an area of interest to vendors who have been perfecting a signature recipe. Based upon parameters such as pH and water activity, further stratification of this category is needed so common food safety processing parameters can be followed to ensure product safety.

Acidified Foods

As described in Chap. 1, many foods are categorized by their pH. Most foods have pH values in the acid range from around 2–6.5. Foods are many times categorized as acid, low-acid, or acidified based on the natural acidity of the product. Foods with a pH at or below 4.6 are considered to be acid foods and those with a pH above 4.6 are called low-acid foods. Acidified foods are low-acid foods that have food grade organic acids or other acidic foods added to produce a final equilibrium pH at or below 4.6. Some examples of acidified foods are pickles, relishes, and salsas.

A pH value of 4.6 has been selected as a limiting factor to control the growth of the anaerobic bacterium, *Clostridium botulinum*. This organism can grow in improperly produced canned, low acid products since they are under a vacuum that is formed when they are filled hot or the product is heated in the container. *C. botulinum* produces a very deadly neurotoxin that causes paralysis and death if not treated. However, *C. botulinum* cannot grow at a pH of 4.6 or lower and, if the bacterium cannot grow, it cannot produce the neurotoxin. Additionally, cooking for an extended period of time at 212 °F (100 °C) will not inactivate spores of *C. botulinum* which can survive any atmospheric cooking and germinate in improperly acidified canned foods. The risk of botulism is the primary reason why specific federal regulations were implemented for acidified foods in the 1970s after outbreaks linked to improperly acidified foods occurred.

There are two federal regulations that govern the safe manufacturing of acidified foods. The US Food and Drug Administration (FDA) regulates products that contain less than 2% cooked meat or poultry or 3% raw meat. Most of the acidified foods sold at farmers markets fall under FDA jurisdiction. The FDA regulations specific to acidified foods can be found in the Code of Federal Regulations, Title 21 Part 114 (21 CFR Part 114). Processors manufacturing a product with >2% cooked meat or poultry or >3% raw meat, should familiarize themselves with the corresponding US Department of Agriculture (USDA) regulations found in 9 CFR Parts 318 and 381.

Federal regulations require establishment of a thermal process to inactivate food safety and spoilage microorganisms and render a product commercially sterile, meaning these microorganisms will not grow under normal storage conditions at room temperature. An individual, known as a processing authority, who is knowledgeable on the thermal processing requirements necessary to achieve commercially sterile product will need to be consulted to establish the appropriate thermal process based on a product's characteristics. Additionally, individuals processing acidified foods for sale must take a one-time education course recognized by the FDA or USDA for processors of these foods, with the most widely known being Better Process Control School.

These regulatory requirements can seem overly burdensome for many vendors who have relatively low volume of acidified foods sales at farmers markets. However, the regulatory oversight provides protection to consumers who can suffer lifetime debilitating ramifications. A single case of botulism is estimated to cost $1.3 million dollars, one hundred times more costly than most other foodborne pathogens [3]. Given this drastic impact to human health, regulations have been placed at the federal level to protect the safety of consumers purchasing manufactured foods, regardless of the manufacturer's size.

While relatively rare, cases of botulism linked to acidified foods for which pH and the accompanying thermal process were not established or controlled for during manufacturing do occur. As an example, two cases of botulism occurred in Ohio in 2014 due to contaminated pine nut basil pesto sold at a farm stand [4]. Upon further investigation, it was determined that the manufacturer, who prepared the foods on behalf of the farm, did not properly establish a thermal process or maintain proper acidification controls, with a finished pH of 5.3 and no other controls in place to limit *C. botulinum* growth such as water activity. This emphasizes the importance of obtaining training specific to manufacturing acidified foods; working with a processing authority to establish an appropriate process; and diligent adherence to this process and proper acidification.

Formulated Acid Foods

Formulated acid foods are very similar to acidified foods in that they rely upon proper acidification to assure control of *C. botulinum*. Therefore, the pH parameters used for acidified foods also apply to formulated acid foods. The primary difference between formulated acid and acidified foods are the amount of low-acid ingredients (pH >4.6). With a formulated acid product, there are very small amounts of low-acid ingredients in the formula. While these products do not require a formal filed process, food manufacturers should work with a processing authority to assure the product fits the definition of a formulated acid food and to determine what heating parameters should be used to assure commercial sterility. They should also have practices in place to monitor processing parameters and product pH.

Low-Acid Canned Foods

Unlike acidified foods, low acid canned foods do not contain sufficient acid to decrease the pH below 4.6 in the product. Products in this category include canned vegetables, soups and some spreads that are meant to be shelf stable. These products must be processed so that a 99.9999999999% reduction of *C. botulinum* spores will occur. In order to achieve this goal, products must be heated at temperatures around 250 °F (121 °C). In order to achieve temperatures in excess of 212 °F (100 °C), equipment that can be operated under pressure, known as a retort, must be used. Retorts differ from pressure cookers in that they are connected to steam, water and air and are fully equipped with controls to provide the appropriate temperature and monitoring devices to record critical parameters such as operating temperature and processing time.

Annually, there are approximately 30–40 cases of botulism that arise when individuals improperly preserve foods in this category. Most commonly the root cause is from water bath preparation, which is far insufficient to inactivate spores of *C. botulinum*. This was the cause of a 2015 botulism outbreak caused by contaminated potatoes that were prepared with a water bath, resulting in 29 cases when the potatoes were used as an ingredient in potato salad served at a potluck meal [5]. Even the use of a pressure canner is insufficient to prepare low-acid canned foods for sale or for use as an ingredient as a processing authority has not been consulted to develop the process, it has not been submitted to the FDA or USDA, and the pressure canner lacks the sufficient controls to assure proper processing temperatures are met.

Jams, Jellies, Fruit Butters and Preserves

Generally speaking, these products pose very low risk if their standards of identity are adhered to, which can be found in 21 CFR Part 150. Federal regulations dictate what ingredients can be added for each product type and certain parameters such as soluble solids, minimum ratios fruit to sweetener, and the necessity to heat in order to solubilize all ingredients. This process provides a product that through heating will inactivate many vegetative foodborne pathogens if present, and the resulting product has a pH and water activity which will not allow for growth of foodborne pathogens.

Typically, when entrepreneurs have issues with these types of products it is because they do not meet the standards of identity put forth in 21 CFR Part 150. For these reasons, it is imperative to assure a formula meets the standards prior to production. Alterations to the formula could result in loss of pH or water activity control, which are imperative to providing shelf-stability and limiting growth of harmful and spoilage microorganisms. Generally speaking, a water activity of 0.85 or below is desirable. In cases where the water activity rises above 0.85, the pH becomes increasingly important and must be closely monitored to assure it does not rise above 4.6. Jellies will not set if they are not properly acidified (pH 3.1–3.2), so monitoring for pH in process can help assure quality as well as safety parameters are met.

Juice and Cider

Juice and cider have been associated with outbreaks since the 1980s and 1990s linked to *Salmonella*, Shiga-toxigenic *E. coli*, *Cryptosporidium*, and Hepatitis A contamination of fresh produce used for juicing [6–9]. Historically, these products were seen as poor vehicles for foodborne pathogens considering they are very acidic. However, several outbreaks proved that foodborne pathogens could survive in this adverse environment and cause illness when consumed.

Based upon these linkages, the FDA has mandated minimum processing requirements to mitigate risks from foodborne pathogens in juice since 2001, known as Juice HACCP (21 CFR Part 120). One of the requirements of Juice HACCP is that processors must implement a processing step which reduces the pertinent foodborne pathogens by 99.999%. However, juice sold directly to consumers in retail outlets meets one of the exemptions for Juice HACCP and therefore many of the vendors selling in farmers markets will not have a primary inactivation step, such as pasteurization, which would reduce the likelihood of foodborne pathogen contamination in these products.

While these vendors are exempt from Juice HACCP, they must adhere to FDA's food labeling regulation in 21 CFR 101.17(g) which requires a warning statement on any juice not processed to inactivate foodborne pathogens which reads: "WARNING: This product has not been pasteurized and, therefore, may contain harmful bacteria that can cause serious illness in children, the elderly, and persons with weakened immune system." Vendors marketing these products should evaluate their ability to institute a processing step which would achieve the Juice HACCP mandate (heat, UV light, high pressure processing) and highly scrutinize their produce suppliers for adherence to GAPs. Unpasteurized juices must be kept refrigerated once packaged and should have a date to reflect proper shelflife, which should not exceed seven days and in some instances may be shorter.

Breads and Cookies

Breads and cookies are yet another group of foods commonly found at the farmers market. These products are routinely seen as non-potentially hazardous (not requiring time/temperature control for safety) given the required baking to achieve desired finished product quality and the low water activity (a_w) of the finished product. As such they are popular among home-based food processors who have cottage food laws in their states permitting sales of these products, as discussed in Chap. 3. These products are rather robust from a microbiological standpoint; however, they can easily be contaminated after baking. It should be noted, that prevention of cross-contamination in a home-based business would be key for cottage food processors. This would rely upon them recognizing training on food safety as a high-priority for their operation. When surveying regulators and food safety educators, the mean response indicated that most cottage food manufacturers did not have the capital to pay to attend food safety training [10].

Beyond the risks caused by microbiological hazards are those posed by undeclared food allergens caused by cross-contact during preparation or the lack of knowledge regarding proper allergen labeling on packages. Harrison et al found that most food safety educators and regulators estimated that <25% of cottage food manufacturers could identify all eight major U.S. allergens, and <40% were aware of mandatory allergen labeling requirements [10]. This is a great concern for products in this category given that many products could contain several of the major eight food allergens, including wheat, eggs, milk, tree nuts, peanuts, and soy. Based upon this information, several food manufacturers operating cottage establishments may not properly identify allergens on their labels, which is the only way for those who are allergic to avoid these items. Additionally, lack of knowledge about food allergens will also most likely result in practices that will allow allergen cross-contact with non-allergen containing foods produced in the same environment. Education to vendors who sell goods in this category at farmers markets would be beneficial to help mitigate risks arising from allergens.

Summary

There is a wide array of products which are further processed for sale at farmers markets. The number of vendors making value-added goods is increasing for numerous reasons, and this interest will result in the need for further resources being devoted to food safety education in the future. Targeted training focusing on the nuances of the product and the processes used in manufacturing will allow vendors to understand where food safety risks occur with a product and how they can mitigate risks in their operation. Additionally, as many vendors are exempt from federal and state regulations, incentivizing education on safe food manufacturing may increase participation in training programs.

References

1. USDA-NASS (2015) Local food marketing practices survey [cited 25 Jan 2016]. https://www.agcensus.usda.gov/Publications/2012/Online_Resources/Local_Food/index.php
2. Wallner-Pendleton EA, Patterson PH, Kariyawasam S, Trampel DW, Denagamage T (2014) On-farm risk factors for *Salmonella* Enteritidis contamination. J Appl Poult Res 23(2):345–352

3. Scharff RL (2012) Economic burden from health losses due to foodborne illness in the United States. J Food Prot 75(1):123–131

4. Burke P, Needham M, Jackson BR, Bokanyi R, St Germain E, Englender SJ (2016) Outbreak of foodborne botulism associated with improperly jarred pesto – Ohio and California, 2014. Morb Mortal Wkly Rep 65(7):175–177

5. McCarty CL, Angelo K, Beer KD, Cibulskas-White K, Quinn K, de Fijter S et al (2015) Large outbreak of botulism associated with a church potluck meal – Ohio, 2015. Morb Mortal Wkly Rep 64(29):802–803

6. Boase J, Lipsky S, Simani P, Smith S, Skilton C, Greenman S et al (1999) Outbreak of *Salmonella* serotype Muenchen infections associated with unpasteurized orange juice – United States and Canada, June 1999 (reprinted from MMWR, 1999, vol 48, pg 582-585). JAMA J Am Med Assoc 282(8):726–728

7. Parish ME (1997) Public health and nonpasteurized fruit juices. Crit Rev Microbiol 23(2):109–119

8. Cook KA, Dobbs TE, Hlady G, Wells JG, Barrett TJ, Puhr ND et al (1998) Outbreak of *Salmonella* serotype hartford infections associated with unpasteurized orange juice. JAMA J Am Med Assoc 280(17):1504–1509

9. Cody SH, Glynn MK, Farrar JA, Cairns KL, Griffin PM, Kobayashi J et al (1999) An outbreak of *Escherichia coli* O157: H7 infection from unpasteurized commercial apple juice. Ann Intern Med 130(3):202–209

10. Harrison JA, Critzer FJ, Harrison MA (2016) Regulatory and food safety knowledge gaps associated with small and very small food businesses as identified by regulators and food safety educators – implications for food safety training. Food Prot Trends 36(6):420–427

Chapter 7
An Overview of Farmers Markets in Canada

Heather Lim and Jeff Farber

Abstract The popularity of farmers markets in Canada has grown immensely over the past 10–15 years, driven by consumer desires for local foods, the interactive experience, and safe and fresh produce. There have been very few outbreaks linked to products purchased or consumed at farmers markets in Canada. A total of six outbreaks were reported; two due to consumption of cheese made from unpasteurized sources, two due to cheese made from pasteurized sources, one due to pork products and one due to unpasteurized fruit cider. No outbreaks have been linked to fresh fruits and vegetables sold at farmers markets in Canada. The few existing microbiological surveys have indicated that fresh fruits and vegetables sold at farmers markets and at retail outlets are comparable in terms of safety, although more surveillance data would allow for a more complete comparison. The few surveys that have been conducted on food handling and temperature control at farmers markets have shown a need for further work. Farmers markets in Canada are regulated at the federal, provincial/territorial and municipal levels. Each province/territory has its own approach to managing potentially hazardous or high risk foods at farmers markets, and may use prohibition or licensing requirements. Farmers market associations at the provincial/territorial level are active participants in the training of market managers and vendors, in partnership with their provincial/territorial governments. Food safety should be increasingly promoted through targeted training of farmers market staff, vendors, and producers.

Keywords Farmers market • Public market • Canada • Food safety • Outbreak • Recall • Province • Territory • Produce • Fruit • Vegetable • Cheese • Cider • Pork

H. Lim
Bureau of Food Surveillance and Science Integration, Health Canada/Government of Canada, 251 Sir Frederick Banting Driveway, A.L.2204D, Ottawa, ON, Canada, K1A 0K9
e-mail: Heather.Lim@hc-sc.gc.ca

J. Farber (✉)
Department of Food Science, University of Guelph,
50 Stone Road East, Guelph, ON, Canada, N1G 2W1
e-mail: jfarber@uoguelph.ca

© Springer International Publishing AG 2017
J.A. Harrison (ed.), *Food Safety for Farmers Markets: A Guide to Enhancing Safety of Local Foods*, Food Microbiology and Food Safety,
DOI 10.1007/978-3-319-66689-1_7

• Unpasteurized • Regulations • Guidelines • *Salmonella* • *E. coli* O157:H7 • *Listeria monocytogenes* • *Campylobacter* • Food handling • Temperature • Training Associations • Economic • Health Canada • Canadian Food Inspection Agency • Public Health Agency of Canada

There has been a recent surge of interest across Canada in farmers markets. In fact, over the past 10 years, the number of farmers markets and their associated economic and social contributions has significantly increased. Some of the drivers for this are consumer interest in buying local products, and the perception that food sold at farmers markets is fresher, of higher quality, and safer than that found at traditional food retailers [1].

Many provinces recently evaluated the economic impact of farmers markets, and the data show the tremendous growth of this sector. In British Columbia, the number of farmers markets increased by 62% from 2006 to 2012 (a total of 159 markets) and the annual economic impact was estimated at $171 million [2]. The economic impact in Ontario in 2008 was estimated at between $641 million and $1.9 billion annually with 15 million shopper visits to 154 markets [1]. The Atlantic provinces have also seen tremendous growth with the number of markets in New Brunswick increasing by approximately 35% over the past 5 years and the number of markets in Nova Scotia tripling since 2004 [3, 4]. Since 2004, the estimated total of direct sales from farmers markets has more than tripled in Alberta, and approximately 72% of households in that province shop at one of the province's 125 farmers markets [5]. In Manitoba, the annual sales by farmers markets almost quadrupled from 2003 to 2008, and the number of vendors increased by 51% [6]. In Québec, the number of farmers markets has grown by almost 30% since 2007 [7].

This chapter provides a snapshot of the current knowledge related to food safety risks from farmers markets in Canada by summarizing known outbreaks and recent recalls, microbiological surveys, and studies of vendor food handling practices. In addition, federal, provincial and territorial regulations and guidelines related to farmers markets, as well as guidelines for specific products sold at farmers markets are summarized. Although not covered in detail in this chapter, it is important to note that provincial farmers market associations play an important role in actively promoting food safety training and safe food practices to their member markets, and local market by-laws may also contain food safety requirements for their vendors.

Outbreaks and Recalls

Six outbreaks were found linked to products sold at farmers markets in Canada from 1994 to 2014 through literature, Internet searches and examination of the Public Health Agency of Canada's *Publically Available International Foodborne Outbreak Database* (Table 7.1). Two of these outbreaks were caused by cheese made from

Table 7.1 Published outbreaks of foodborne illnesses associated with farmers markets in Canada

Date	Food	Organism	Location	Number of cases	Comments	Reference
September 1994	Soft cheese, unpasteurized sources	*Salmonella* Berta	Waterloo, ON	82	Cross-contamination during processing	Ellis et al. [8]
February 2002	Soft ripened cheese, pasteurized sources	*L. monocytogenes*	Vancouver, BC	49	Post-pasteurization contamination	McIntyre et al. [10]
August–September 2002	Soft ripened cheese, pasteurized sources	*L. monocytogenes*	Vancouver Island, BC	86	Post-pasteurization contamination from water system. A different processor than the Feb 2002 outbreak	McIntyre et al. [10]
December 2002	Gouda cheese, unpasteurized sources	*E. coli* O157:H7	Edmonton, AB	13	Source not determined. Cheese was in compliance with 60-day aging and microbiological requirements.	Honish et al. [9]
August 2006	Pork products	*S.* Typhimurium PT 170	PEI	5		PHAC [11]
October–November 2014	Apple cider, unpasteurized	*E. coli* O157:H7	Waterloo, Ontario	5	Of the five cases, three were under 16 years of age	PHO and CFIA [13, 18]

unpasteurized sources, one due to *Salmonella* Berta with 82 cases in 1994 in Waterloo, Ontario and the other due to *E. coli* O157:H7 with 13 cases in 2002 in Edmonton, Alberta [8, 9]. Two outbreaks of *Listeria monocytogenes* from soft ripened cheese made from pasteurized sources occurred in the same year and province (2002, British Columbia), but from different manufacturers [10]. There were 49 and 86 cases attributed to these two cheese outbreaks from pasteurized sources. One outbreak of five cases of *Salmonella* Typhimurium PT 170 in Prince Edward Island (PEI) was attributed to "pork products", with no further information [11]. One outbreak in 2014 caused by *E. coli* O157:H7 from unpasteurized apple cider was reported as causing five cases in Waterloo, Ontario [12, 13]. These six outbreaks occurred at locations across Canada. No published outbreaks from farmers markets were found prior to 1994. It is possible that outbreaks linked to farmers markets are less likely to be identified than those from large retailers due to the smaller volume of products being purchased.

Four other outbreaks linked to fairs/festivals and one outbreak linked to a charity event provide some information about areas of potential risk for farmers markets. Thirty-five cases associated with *Shigella sonnei* infection were reported due to the consumption of fresh parsley, which was a component of a smoked salmon and pasta dish at a fair in Ontario in 1998 [14]. Furthermore, three cases of *E. coli* O157:H7 were caused by the consumption of unpasteurized milk at an Agricultural Exhibition in Lévis, Québec in 1994 [15], and in Manitoba in 2010, 40 cases of *E. coli* O157:H7 were linked to cross-contaminated fruit compote or other foods served at a local festival [16]. In this latter case, the kitchen had been inspected, and staff had been trained. Watermelon jelly sold at a table outside a grocery store and at charity booths caused one case of botulism in British Columbia in 2010 [17]. In addition, a large outbreak due to *Staphylococcus aureus* occurred in Toronto in 2013, when 223 people became sick after eating improperly refrigerated maple bacon jam [15]. Examples of recalled products sold at farmers markets include jarred liver pâté due to *Clostridium botulinum* concerns and unpasteurized fruit cider due to *Salmonella* concerns [18, 19].

Some of the trends seen in these outbreaks were that illnesses were associated with high-risk foods (e.g., cheese from unpasteurized sources and unpasteurized juice and milk) and improperly canned foods. In addition, there were issues with maintaining refrigeration temperature and cross-contamination. Some of the ways these risks are currently being addressed are through Federal and Provincial Acts and Regulations, guidance documents for cheese made from unpasteurized sources and unpasteurized fruit juice, the prohibition of the sale of unpasteurized milk by federal and some provincial regulations, and provincial guidance for inspectors, market managers and vendors.

Potentially Hazardous Foods and Regulatory Approaches

The provinces and territories in Canada have different ways of addressing the risk posed by potentially hazardous foods sold at farmers markets. Most provinces have developed categories of foods based on food safety risk, although the specific foods identified, the various terms used and the requirements for each of the categories differs from province to province. In some cases, foods considered a high risk such as unpasteurized milk are banned from sale (federal Food and Drug Regulations and some provincial regulations). Other examples of foods that are prohibited from sale at farmers markets are ungraded eggs and uninspected meat in Ontario, and in Newfoundland and Labrador, meat from unlicensed abattoirs and home bottled (canned or processed) low-acid products.

Another category of food identified in provincial guidelines is "potentially hazardous foods" or "higher risk foods", with different terminology used in different provinces. These foods are considered potentially hazardous, but are allowed for sale in most provinces. Each province has their own list, but in general, examples include ready-to-eat foods that require refrigeration, fresh meat, sausages, foods containing eggs or dairy products, fish and shellfish and home canned low-acid vegetables. Provinces have different ways of addressing the risk posed by potentially hazardous foods sold at farmers markets, often relying on required licensing or permits. One example of provincial requirements for potentially hazardous foods can be found in the 2015 British Columbia (BC) Centre for Disease Control (CDC) Guidelines where it specifies that higher risk foods are not permitted for home preparation, vendors must submit an application to the local health authority to sell them, they must be maintained at a temperature of 4 °C (39.2 °F) or colder, and in the case of raw poultry, meat and fish, the products must be kept frozen [20]. In New Brunswick, foods are categorized into four "classes", with the higher risk classes having specific licensing requirements [21]. For potentially hazardous foods, the Newfoundland and Labrador guide specifies that they must be maintained at safe temperatures [22]. In Québec, there are two categories of permits that are required for specific types of foods sold at farmers markets [23]. Further details on each province's requirements are provided in Table 7.2 and the section on Food Safety Oversight.

Comprehensive surveys on the microbiological safety, food handling and actual storage temperatures of many of these potentially hazardous foods appear to be lacking for farmers markets in Canada.

Cheese

Four outbreaks linked to cheese sold at farmers markets in Canada have been reported (Table 7.1). Two of these outbreaks involved cheese made from pasteurized sources and two involved cheese from unpasteurized sources. The two outbreaks linked to soft ripened cheese made from pasteurized sources both occurred in 2002 in British Columbia due to post-pasteurization contamination with *L. monocytogenes* serotype 4b [10]. The cheeses responsible for the outbreaks were made by two different producers, sold at farmers markets in separate locations, and the PFGE patterns of the *L. monocytogenes* isolates were different. The first outbreak occurred in February of 2002 in Vancouver, British Columbia and 49 cases were confirmed.

Table 7.2 Provincial guidelines and/or regulations dealing with farmers markets (FMs) in Canada

Province	Legislation/Regulations	Guidelines	References	Comments
Newfoundland and Labrador (Public Market)	The legislated responsibility falls under the provincial *Food and Drug Act* and subsequent *Food Premises Regulations*	The Public Market Guidelines, September 2011, define a *Public Market* as a venue where organized groups of vendors gather on a regular basis in a common location to market food products for which they are directly responsible. The larger scope of high-risk, ready-to-eat (RTE) food products differentiates this market from traditional Farmers Markets (FM)	Public Market Guidelines (2011) https:// static1.squarespace.com/ static/54d9128be4b0de7874ec9a82/t/562a e6c4e0411e298c1ca8/1445652164881/ Public_Market-Guidelines_2012.pdf	It is the role of Environmental Health Officers within Service Newfoundland and Labrador to inspect all commercial food premises where food is prepared or served to the public
	These regulations apply to all commercial food service in Newfoundland and Labrador, regardless of its origin	Handbooks for Market Organizers and Vendors have been developed that include food temperature monitoring sheets, start-up checklists, food safety checklists and food safety resources	Farmers Market Food Safety: Market Organizer Handbook (2011) http://www. foodsecuritynews.com/Publications/ Farmers_Market_Food_Safety_ Organizer_Handbook.pdf	Foods are categorized based on risk into Schedule A, B or C. Vendors are required to obtain a license to sell Schedule A and B products.
			Farmers Market Food Safety: Vendor Handbook (2011) http://www. foodsecuritynews.com/Publications/ Farmers_Market_Food_Safety_Food_ Vendor_Handbook.pdf	Products in Schedule C, such as home bottled meat or fish, meat or poultry from unlicensed abattoirs or non-acidified canned foods are not permitted for sale at farmers markets

| Prince Edward Island | The legislated responsibility is the Food Premises Regulations, under the authority of the *Public Health Act* | Guidelines also exist for Food Vendors at Farmers Markets | Food Premises Regulations https://www.princeedwardisland.ca/sites/default/files/legislation/p30-1-02.pdf Prince Edward Island—The Department of Health and Wellness https://www.princeedwardisland.ca/en/information/health-and-wellness/food-premises-program | Under the direction of the Chief Public Health Officer, the Dept. of Health and Wellness (Environmental Health Section) enforces the minimum standards as set out in the regulations. - FM managers must apply each year to Environmental Health for a Type 5 Approval Certificate - Food vendors must apply to Environment Health to receive an approval certificate - Only those food vendors who prepare for sale or sell RTE foods require an approval certificate to operate |

(continued)

Table 7.2 (continued)

Province	Legislation/Regulations	Guidelines	References	Comments
Nova Scotia (Public Market)	There does not appear to be any regulations specifically pertaining to farmers markets	There exists Food Safety Guidelines for Public Markets, which are designed to assist public market vendors only	The Food Safety Guidelines for Public Markets (2016) http://novascotia.ca/agri/documents/food-safety/publicmarketguide.pdf	As per the guidelines, a *Public Market* is a venue where organized groups of three or more vendors gather on a regular basis in a common location to market products for which they are directly responsible. The sale of Schedule A and Schedule B food products at these venues is restricted to no more than two business days per week. This includes the operation of Farmers Markets and Flea Markets
	It is the role of Nova Scotia Environment—Environmental Health and Food Safety Division to inspect all premises, whether permanent or temporary, where food is prepared or served to the public. The legislated responsibility for this falls under the *Nova Scotia Heath Protection Act and associated regulations*		Fact sheets and publications on food safety and public markets http://novascotia.ca/agri/programs-and-services/food-protection/factsheets-publications/	Permits must be obtained for Schedule A (lists foods that are classed as potentially hazardous foods) and Schedule B (lists food products that are *not considered* potentially hazardous food). Schedule C—foods that present significant potential risk to the public health when conditions are compromised or may contravene existing legislation—these are not permitted for sale to the public in a Public Market
				Website, training and resources available at: http://farmersmarketsnovascotia.ca

New Brunswick (Public Market)	The Food Premises Regulation came into effect November 20, 2009, under the New Brunswick *Public Health Act*	The New Brunswick Guidelines for Food Premises at Public Markets, April 1, 2016, provides licensing requirements, information and guidance to public market operators and their licensed food vendors on safe operations, from a food safety perspective	The New Brunswick Guidelines for Food Premises at Public Markets (2016). http://www2.gnb.ca/content/dam/gnb/Departments/h-s/pdf/en/HealthyEnvironments/Food/NBMarketGuidelines_E.pdf	Foods not permitted at Public Markets are as follows:
	Under the legislation, some vendors require a license dependent upon the type of food(s) the vendor is preparing and/or selling. - Class 3: potentially hazardous foods. - Class 4: food is prepared or processed for sale or consumption on or off the premises, but is not distributed wholesale.			• All milk and milk products made with raw milk unless they are from a provincially licensed Class 5 Dairy Plant or federally registered establishment • Low-acid canned/bottled foods not produced in a licensed facility
	Class 3 and Class 4 food premises are required to have a food premises license. Each class has specific licensing requirements			• Bottled meat, canned fish, smoked fish, shellfish and seafood not produced in a licensed facility • Meat and poultry or meat and poultry products from animals that have not been slaughtered at a licensed or registered establishment. • Wild foraged mushrooms • Any food product that requires Class 5 operations/processes at the market
	The regulation defines *public market*, not farmers market. Public Market means a venue where a group of vendors sets up on a regular basis in a common location to sell food products, and includes a farmers market and a flea market			If unpasteurized juice or cider is being sold, it must be labelled as "UNPASTEURIZED"

(continued)

Table 7.2 (continued)

Province	Legislation/Regulations	Guidelines	References	Comments
Québec (Public Market)	The Food Products Act (LRQ, chapter P-29) and regulations on Food and on Fruits and Vegetables (P-29, r1 and r3)	A guide was prepared by MAPAQ for agricultural operators (farmers) in public markets with information on: - licenses required for various situations - compliance with regulations related to training in food hygiene and safety - information on product origin - labeling	Document d'information, Marchés publics exigences de commercialisation (2013) http://www.mapaq.gouv.qc.ca/fr/Publications/Marche_public.pdf	A website on Québec Public Markets exists for local communities, producers and artisans. http://www.ampq.ca/
	According to the types of foods offered for sale (fresh whole foods, processed foods, maple, honey or eggs) and according to the categories of farmers or agricultural producers, various types of licenses (retail permits "keep hot or cold" or "general preparation") may be required.	A guide on Good hygienic practices in the public market provides information on a number of topics, including labelling, safe food storage temperatures, cleaning and sanitizing, safe food handling, contact surfaces, and packaging	Guide de bonnes pratiques d'hygiène pour les marchés publics alimentaires (2009)	

Ontario	The Food Premises Regulation O. Reg. 562/90, under the *Health Protection and Promotion Act* (HPPA)	Operational guidelines exist, entitled "Common Approaches for Farmers Markets and Exempted Special Events", developed by the Association of Supervisors of Public Health Inspectors of Ontario (revised May 2012)	Common Approaches for Farmers Markets & Exempted Special Events (2012) http://www.farmersmarketsontario.com/DocMgmt%5CFood%20Safety%5CHealth%20Unit%20Guidelines%5CASPHIO-CommonApproachesGuidelinesRevisedMay2012.pdf	Ontario website on farmers markets exists for managers, vendors and shoppers. http://www.farmersmarketsontario.com/
	The Food Premises Regulation defines farmers markets as a central location at which a group of persons who operate stalls or other food premises meet to sell or offer for sell to consumers products that include, without being restricted to, farm products, baked goods and preserved foods, and at which the majority of the persons operating the stalls or other food premises are producers of farm products who are primarily selling or offering for sale their own products http://www.ontario.ca/laws/regulation/900562			Regulatory requirements for fruit, vegetables, nuts, honey and maple do apply to produce at farmers markets
	Farmers markets are exempt from the *Food Premises Regulation* under the Minister of Health and Long-Term Care, HPPA only if greater than 50% of vendors are producers of farm products, who primarily sell their own products. Certain provisions of the Act still apply			*Livestock and Livestock Product ACT* prohibits the sale of ungraded eggs beyond the farm gate

(continued)

Table 7.2 (continued)

Province	Legislation/Regulations	Guidelines	References	Comments
Manitoba	*No* specific regulations for farmers markets	Farmers Market Guideline, June 2009, defines a Farmers Market as a short-term operation for the sale of produce and prepared food products under the direction of a designated operator	Farmers Market Guideline (2009) https://www.gov.mb.ca/health/publichealth/environmentalhealth/protection/docs/farmers_market.pdf	Additional items may be required by the Public Health Inspector pursuant to the Food and Food Handling Establishments regulation—MR 339/88R (*The Public Health Act*). Operators must have a valid permit issued by the Public Health Inspector
	There are plans to write farmers markets into the regulations			Potentially hazardous food must be prepared at an approved Food Handling Establishment and conform to other applicable regulations
Saskatchewan	*No* specific regulation exists for FMs	A Farmers Market Technical Guideline, July 15, 1998, is intended to be used by industry as a guide and by public health inspectors as a reference	The Food Safety Regulations apply to food facilities such as restaurants, and food processing operations. http://www.qp.gov.sk.ca/documents/English/Regulations/Regulations/P37-1R12.pdf	Farmers Market vendors (or employees) are not required, but are encouraged, to have food safety training
	However, as per the *Farmers Market Technical Guideline, July 15, 1998*, the Food Safety Regulations are considered the reference regulation	*Farmers Market*—is a market where rental or allotted space is used to sell items or goods that are made, baked or grown locally and where at least 50% of the spaces are designated for the sale of locally grown produce for at least six months of the year		

Alberta	Part 3 of Alberta's Food Regulation, pursuant to the Public Health Act, sets out requirements for the market itself and the vendors (stall holders)	Alberta Health Services has published a guideline on food safety responsibilities and requirements for Market Managers and Vendors	Alberta Approved Farmers Market Program Guidelines (2015) http://www1.agric.gov.ab.ca/$Department/deptdocs.nsf/all/apa2577	Farmers markets must have Alberta Agriculture and Forestry (AAF) recognition, before Health recognizes them as a farmers market
	Food regulation references a "farmers market permit". Farmers markets must hold a Food Handling Permit issued by Alberta Health Services, the single authority in Alberta which inspects and permits food establishments, including farmers markets		Guidelines for Public Markets Managers and Vendors (2014) http://www.albertahealthservices.ca/assets/wf/eph/wf-eh-guidelines-for-public-market-managers-and-vendors.pdf	
			Farmers Market Home Study Course (2011) http://www.albertahealthservices.ca/assets/wf/eph/wf-eh-home-study-farmers-market.pdf	
	"Farmers Market" means a food establishment whose proposed operation has been approved by the Minister responsible for Agriculture as an approved farmers market program under the administration of that Minister's department		Alberta Food Regulation (2006) http://www.qp.alberta.ca/1266.cfm?page=2006_031.cfm&leg_type=Regs&isbncln=9780779785742	*Training* Farmers Market Home Study Course—developed by the Environmental Public Health to help market managers and vendors set-up and operate an Alberta approved Farmers Market in a sanitary manner. The course is mandatory if the Food Safety training under section 31 of the Food Regulation was not completed

(continued)

Table 7.2 (continued)

Province	Legislation/Regulations	Guidelines	References	Comments
British Columbia	Environmental Health Officers inspect farmers markets and operate under the: 1) Public Health Act 2) Food Safety Act 3) Food Premises Regulation	A guideline published by the BC Centre for Disease Control provides recommendations for the preparation and display of food intended for sale at temporary food markets	Temporary Food Markets, Guideline for the Sale of Foods at Temporary Food Markets (2016) http://www.bcfarmersmarket.org/resources/subpage/health-and-safety	Low-risk foods are acceptable to be made at the vendor's home vs. high-risk foods, which need to be made in a commercial kitchen or approved facility. *Training* 1. "Marketsafe" for vendors producing low-risk foods is highly recommended 2. FOODSAFE level 1 is required for vendors producing high-risk foods
	The Health Authority has discretion to remove any food sold at a temporary food market considered to be a health hazard as defined in Section 1 of the Public Health Act, or is contaminated as per Section 3 of the Food Safety Act			Website of provincial association with information and training on food safety http://www.bcfarmersmarket.org/resources/subpage/health-and-safety
Nunavut	*No specific regulation or guideline for farmers markets*	No		
	Eating and Drinking Place regulations exist under the Public Health Act that deals with food safety			

Northwest Territories	*No* specific regulations for Farmers Markets exist; however, regulations are in the development stages	Guidelines exist for Operating Temporary Food Service Establishments (Northwest Territories Health and Social Services, not dated) *The guidelines define:* - Temporary food service establishments - Food establishment The guidelines also outline approval to operate a Temporary Food establishment, which must be obtained from the Environmental Health department, for each temporary food concession during a public event. In addition, there is information on food source and protection, food temperatures, food serving and handling, handwashing, etc.	Currently, there are not many farmers markets in the NWT https://www.justice.gov.nt.ca/en/files/legislation/public-health/public-health.r8.pdf
Yukon	*No* specific regulations or guidelines for farmers markets *Currently guided by the:* 1. Agriculture Products Act 2. Public Health Act	No	http://www.gov.yk.ca/legislation/legislation/page_a.html

Spray culture solution bottles used by the cheesemaker were concluded as the likely source of the cheese contamination. The second outbreak, with 86 confirmed cases, began in August of 2002 on Vancouver Island, British Columbia. The investigation concluded that the cheesemaker's water source was being contaminated by wild birds and the UV disinfection was ineffective. When lukewarm water, instead of the usual hot water, from the contaminated source was used to wash a batch of the cheese curds, *L. monocytogenes* was introduced into the cheese [10].

Although the Canadian Federal *Food and Drug Regulations* prohibit the sale of unpasteurized milk, the sale of cheese from unpasteurized sources is permitted if the product is aged for a minimum of 60 days at 2 °C (35.6 °F) or more and the end product meets microbiological requirements for *E. coli* and *Staphylococcus aureus* (Food and Drug Regulations B.08.030, B.08.043, B.08.044 and B.08.048(2)) [24]. In addition, section 4 of the *Food and Drugs Act* prohibits the sale of food that has in or on it any poisonous or harmful substance (e.g., pathogens) [25]. In 1994, soft cheese from unpasteurized sources produced on a farm in Waterloo, Ontario caused 82 cases of salmonellosis [8]. The investigation revealed that cross-contamination by chicken carcasses during the cheese production led to the presence of *Salmonella* Berta in the cheese [8]. The second outbreak from unpasteurized sources which occurred in 2002 in Edmonton, Alberta and caused 13 confirmed illnesses with *E. coli* O157:H7, was associated with the consumption of Gouda cheese [9]. The cheese was in compliance with the regulatory requirements for a 60-day aging period and microbiological testing. Investigation of the plant and manufacturing practices could not determine the source of the *E. coli* O157:H7 in the cheese [9]. Health Canada recently published voluntary guidance for manufacturers on improving the safety of soft and semi-soft cheese made from unpasteurized milk, as well as a joint Health Canada/FDA quantitative assessment on the risks of listeriosis associated with the consumption of soft-ripened cheese from unpasteurized sources [26, 27].

Despite these outbreaks, there does not appear to be any published microbiological surveys of cheese sold at farmers markets in Canada. A study published in 2004 observed the food safety practices of 17 cheese vendors at farmers markets in 5 urban regions of South-Western Ontario [28]. The authors noted that overall the food safety practices were good, however, a few areas for improvement were identified. While most of the vendors sold pre-packaged cheese, 41% (7/17) portioned the cheese at the market. A number of vendors (47%) had problems with maintaining proper refrigeration temperatures (5 °C or less; 41 °F or less), and cheese sold by two of the vendors had no temperature control. In terms of handwashing, 24% of vendors did not have sinks available, 88% of handlers did not wash their hands before or after serving customers, and 24% of vendors handled the cheese directly with their hands. Four of the vendors stored cheese next to raw meat products. In the four cheese outbreaks previously discussed, the cheese was contaminated before coming to market and it is unknown to what extent, if any, the food safety practices of the vendors impacted the number or severity of illnesses. Studies on the microbiological quality of cheeses sold at farmers markets and food handling practices of vendors are warranted from different regions and types of markets across the country to more fully understand the potential risks associated with these products.

Unpasteurized Fruit Juice/Cider

Although there have been six documented outbreaks in Canada linked to unpasteurized fruit juice or cider since 1974 [13, 29], only one in 2014 was from a product purchased at a farmers market [12, 13]. Public Health officials recommend that juice or cider be pasteurized, especially before giving to vulnerable populations such as children and the elderly. In a recent outbreak in Waterloo, Ontario, three of the five cases were children, and in past outbreaks, where demographic information was available, a disproportionately high percentage of the cases were children (50–100% of cases) [13, 30–32].

The current Health Canada policy has three components: a Code of Practice with recommendations for the safe production and distribution of unpasteurized juice/cider [33]; voluntary labeling with the word "unpasteurized" to allow for consumer choice; and the production of educational material [34]. At farmers markets, packaged cider and free samples may or may not be labeled as 'unpasteurized' and are readily consumed by the young and old alike. More could be done to promote consumer education surrounding these products at farmers markets and/or recommend that all unpasteurized fruit juice or cider be labeled appropriately, indicating whether or not the product has been pasteurized.

Fresh Fruits and Vegetables

Purchasing fresh fruits and vegetables is often cited as the number one reason to visit a farmers market [1, 6]. No published outbreaks were found linked to the sale or consumption of fresh fruits and vegetables from farmers markets in Canada. However, fresh fruits and vegetables have been the cause of outbreaks from farmers markets in the U.S. and at retail in Canada, showing that these products have the potential to cause foodborne illness. In the U.S., several outbreaks have been linked to fresh and fresh-cut fruits and vegetables sold at farmers markets including raw bagged peas, strawberries, and fresh-cut samples of cantaloupe [35, 36]. Fresh and fresh-cut fruits and vegetables sold at retail have been linked to 27 outbreaks in Canada from 2001 to 2009, 48% of which have leafy green vegetables as either a confirmed or suspected vehicle [37]. Because of the smaller volume of fresh fruits and vegetables sold at farmers markets, it is possible that illnesses caused by these products are more likely to be under-reported, as compared to those illnesses caused by fresh fruits and vegetables sold at retail. Many of the potential hazards are similar for both produce sold at retail and at farmers markets. For leafy green vegetables, some of the identified risk factors include manure management, irrigation and post-harvest water quality, presence of animals in the field, and growth of pathogens after cutting and during storage and transport [38].

The Codex Alimentarius "Code of hygienic practice for fresh fruits and vegetables" and "Recommended international code of practice general principles of food hygiene" are international guidelines which address good agricultural and manufacturing practices to reduce the risk of fresh fruits and vegetables becoming contaminated with pathogens, as well as chemical and physical hazards [39, 40]. Studies conducted by the British Columbia Ministry of Environment examining the compliance of farmers to good agricultural practices (GAPs) relating to manure storage and composting, irrigation and separation of animals from waterways, found that hobby farms have consistently lower compliance than commercial farms [41]. The main reasons given were a lack of awareness of best practices and regulatory requirements. Another study of small-scale and organic producers in Ontario found that participants reported a high use of most GAPs, even though only 26% of respondents participated in an on-farm food safety program [42]. An area for improvement that was identified is the disinfection of produce wash water and livestock and poultry drinking water [42]. Similar concerns were found for small to medium-sized farms in a U.S. study where 16% of survey respondents used untested well water for washing produce and of those farmers that used manure to fertilize crops, 36% used improperly composted manure [43]. A study of produce vendors in farmers markets in North Carolina, U.S., showed that 85% were unaware of GAPs or unresponsive when prompted and of those that were aware, many believed GAPs were not applicable to farmers markets [44]. While fresh fruits and vegetables sold at farmers markets do enjoy a history of safety, the outbreaks at farmers markets in the U.S. and at retail in Canada should encourage the awareness of the risks associated with these products and promote the implementation of best management practices on-farm and at market.

Four microbiological surveys which sampled produce at Canadian farmers markets in four different locations have been published, allowing a limited comparison of produce from farmers markets, organic produce, produce sold at supermarkets and imported produce. Most of them focused on leafy green vegetables which were ranked as the highest priority based on microbiological risk, according to a 2007 expert meeting convened by the FAO and WHO [45]. The studies collected samples from local farmers markets around Vancouver, British Columbia (indicator organisms only); the province of Alberta (indicators and pathogens); south-central and south-western Ontario (indicators and pathogens); and, Ottawa, Ontario (thermotolerant *Campylobacter* only) [46–49]. Overall, the percentage of samples testing positive for generic *E. coli* for leafy greens and herbs ranged from 0 to 27.1%, with the counts ranging from undetectable to 4.2 log CFU/g. Lettuce samples from Ontario had a lower percentage of positive samples (5.7%, n = 530) for generic *E. coli* than those from British Columbia (13%, n = 68) and Alberta (18%, n = 128). Further analysis of the generic *E. coli* isolated from the British Columbia study showed that 97% possessed resistance to at least one antimicrobial.

Two of the studies compared generic *E. coli* counts between organically- and conventionally-grown produce and neither found a significant difference between the two production methods, although, in one of the studies, the authors questioned whether this is due to the absence of any real differences or to low numbers of positive samples [47, 48]. The study from Ontario also compared produce from retail (distribution centres) and farmers markets and found no significant difference between the generic *E. coli* rates [48].

The types of pathogens tested varied between studies; overall, one spinach sample was positive for *Cryptosporidium* (n = 59) and two were positive for *Salmonella* Schwarzengrund (one tomato, n = 141, and one organic leaf lettuce, n = 112). The study by Parks and Sanders sampled vegetables from farmers markets (during the summer) and supermarkets (during winter and summer) in Ottawa, Ontario and tested for thermotolerant *Campylobacter* spp. [49]. Their results showed a higher level of *Campylobacter* spp. in produce from farmers markets (1.7%, 9/533) than produce from supermarkets (0%, 0/1,031). Spinach and lettuce had the highest rates of positive samples at 3.3% (2/60) and 3.1% (2/67), respectively. Farmers market samples were negative for *Campylobacter* spp. after being thoroughly washed with chlorinated water. The authors pointed out that *Campylobacter* spp. are sensitive to oxygen and other environmental stresses, and that rapid die off would likely occur during transportation and storage. Farmers market produce may be fresher and so there may not be as much die-off of *Campylobacter* spp. in products that are sold and consumed quickly. There is the notion among both consumers and producers that the food sold at farmers markets and, in general, food produced locally, is superior in terms of quality and safety to imported products [1, 42]. It is important that consumers understand that safe food-handling practices are critical regardless of where food is purchased.

There are several data sources that provide a more comprehensive picture of the microbiological quality of domestic and imported fruits and vegetables at supermarkets and a limited comparison can be made to that of farmers markets. The Canadian Food Inspection Agency has published several recent reports on their targeted surveillance programs for fresh fruits and vegetables. According to three large, national surveys, the vast majority (>99%) of imported and domestic fruits and vegetables sold at retail had no detectable pathogens and had acceptable *E. coli* results (based on Health Canada's criteria for generic *E. coli*: n = 5, c = 2, m = 100 CFU/g, M = 1000 CFU/g where n is the number of samples, c is the maximum number permitted in the marginal range, m is the lower limit of the marginal range and M is the upper limit of the marginal range) [50–52]. Based on the limited surveillance data for farmers markets in Canada, the microbiological quality of fruits and vegetables from both farmers markets and retail appears to be comparable. However, outbreaks at retail in Canada and in farmers markets in the U.S. show that vigilance in following GAPs and food handling and storage guidelines is warranted.

Food Handling Practices of Farmers Market Vendors in Canada

Very few surveys have been published on the safe handling practices of vendors at farmers markets in Canada. One study, published in 2014, observed a total of 21 vendors from seven farmers markets located in either metro-Vancouver or Northern British Columbia [53]. Compliance was observed for the following five categories: general compliance, handwashing, samples for tasting, temperature control, and preserved items. Similar to the cheese vendor survey cited above, issues with handwashing and temperature control were highlighted. Handwashing stations were available to 63.6% (7/11) of the observed vendors, with only 45.5% having complete handwashing supplies. None (0/11) of the vendors serving samples were observed to wash their hands during the observation period (6 hours), and six of these vendors portioned the samples at the market. None of the four vendors that wore gloves used them properly, and 90.9% of vendors handled both food and money. In terms of temperature control, only one of ten vendors had a thermometer, four had food temperatures above 4 °C (39.2 °F) and one had no method to cool the food. As the authors of the study point out, these poor food handling behaviors were observed despite 'good' food safety knowledge of market vendors and the presence of handwashing stations. The observed handwashing behaviors and issues with temperature control in these two Canadian surveys (including the cheese vendor survey) align with the results from a study of employees at farmers markets in Indiana, U.S., where handwashing was only observed twice during a total of 417 transactions that required handwashing [54]. They also align with a study of North Carolina Farmers Market vendors where only one thermometer was seen used after observing 168 vendors, and temperature abuse of potentially hazardous foods occurred [44]. In the U.S. studies, handwashing stations were unavailable to 50% and 19% of vendors, respectively. Education in this area is definitely warranted. Safe food handling behaviors and a strong food safety culture need to be actively promoted by market managers.

Food Safety Oversight of Farmers Markets in Canada

Farmers markets are subject to the Federal *Food and Drugs Act and Regulations*, *Consumer Packaging and Labeling Act*, and the *Canada Agricultural Products Act*. In Section 4 of the Food and Drugs Act, it states "(1) No person shall sell an article of food that

(a) has in or on it any poisonous or harmful substance;
(b) is unfit for human consumption;
(c) consists in whole or in part of any filthy, putrid, disgusting, rotten, decomposed or diseased animal or vegetable substance;
(d) is adulterated; or
(e) was manufactured, prepared, preserved, packaged or stored under unsanitary conditions."

Also, as previously mentioned, the sale of unpasteurized milk is prohibited in Canada.

In addition to the federal requirements, provinces and territories may have their own regulations specific to farmers markets, as well as those of municipalities and local public health units. Table 7.2 summarizes the provincial and territorial regulatory status of farmers markets in Canada as well as available guidelines.

All provinces and territories appear to have food safety guidelines in place that are applicable to farmers markets or public markets, except for Nunavut and Yukon. In addition, some provinces were found to have legislation and regulations specifically related to farmers markets or public markets. The following is a short synopsis of what some of the provinces have in their regulations or guidance documents, as pertaining to farmers markets.

Newfoundland and Labrador The legislated responsibility falls under Newfoundland and Labrador *Food and Drug Act*, and *Food Premises Regulations* and the public market guidelines. Newfoundland and Labrador uses the term "Public Market", as it represents a broader scope of operation that includes a wide variety of food products.

Various government departments have recognized the need to apply a level of control that will ensure consumer protection and encourage food security while providing entrepreneurial opportunities. Food sales for Public Markets are classified into three groups:

- Schedule A foods, are hazardous foods that normally are processed or could easily spoil and lead to illness.
- Schedule B foods are naturally less hazardous and require fewer procedures to ensure safety.
- Schedule C foods are foods that cannot be sold in public markets because they are considered too hazardous.

The Public Market Guideline also provides information related to e.g., source of ingredients, licensing requirements, general food hygiene, food labelling and allergens, pesticides, eggs, etc. Additionally, on-line there is a "Farmers Market Food Safety: Market Organizer Handbook—A Resource for Farmers Market Organizers in Newfoundland & Labrador" which provides information for organizers of farmers markets to help ensure food safe practices and compliance with provincial food safety regulations. There is also a handbook which provides an overview of food safety, including regulation, licensing requirements and vendor-specific responsibilities for farmers market vendors [55].

Nova Scotia There are Food Safety Guidelines for Public Markets that cover among other things, policy, definitions, potentially hazardous foods and public market permits [56]. It is the role of the Nova Scotia Environment—Environmental Health and Food Safety Division to inspect all premises, whether permanent or temporary, where food is prepared or served to the public. Information is also provided in Schedule A (potentially hazardous foods) and Schedule B (potentially non-hazardous foods). There is also a quick reference guide for products and the conditions for sale and food handling. Some of the products listed include fruits, vegetables, honey, meat, fish, poultry, dairy products, baked goods, eggs, cold and frozen drinks and processed foods, including cabbage rolls, cured ham or bacon and pickled eggs. Additionally, there are on-line fact sheets for Public Markets that provide information on food safety.

Prince Edward Island Under the *Public Health Act*, the *Food Premises Regulations* contain very detailed requirements, which include two Appendices [57]. The regulation covers requirements such as licensing; design and construction, staff hygiene and temperature requirements. Appendix I focuses on cleaning and sanitizing, and Appendix II focuses on temperature and time control. A safe internal cooking temperature chart is also provided.

New Brunswick Under the *Public Health Act*, the *Food Premises Regulation* covers among other things, standards for food premises, control of food hazards, e.g., handling of potentially hazardous foods, maintenance and sanitization, hygiene and diseases, as well as training [21]. New Brunswick uses the term "Public Market". Guidelines have been developed by the New Brunswick Department of Health and Health Protection Branch entitled "The New Brunswick Guidelines for Food Premises at Public Markets" [21]. These guidelines provide licensing requirements, information and guidance to public market operators and their licensed food vendors on safe operation from a food safety perspective. They outline related processes and requirements for licensed public market food premises that are within New Brunswick Department of Health's legislated authority.

Québec There are two guides published by the Ministère de l'Agriculture, des Pêcheries et de l'Alimentation du Québec (MAPAQ) that pertain to public markets. One outlines marketing requirements for public markets, and the other is a guide for good hygienic practices [23, 58]. Vendors are required to obtain permits from MAPAQ to sell prepared foods, foods requiring warm or cold storage temperatures, and certain products such as those containing maple, honey or egg. Other requirements are that meats sold must come from a federally or provincially inspected abattoir, and only graded eggs are permitted to be sold at farmers markets [23]. The "Guide de bonnes pratiques d'hygiène pour les marchés publics alimentaires" (2009) provides information on a number of pertinent topics, including labeling, safe storage temperatures, cleaning and sanitizing, safe food handling, contact surfaces, and packaging [58].

Ontario Under the *Health Protection and Promotion Act*, the *Food Premises Regulation* defines farmers markets. Farmers markets are exempt from the *Food Premises Regulation* only if greater than 50% of vendors are producers of farm products who primarily sell their own products. Guidelines have been developed entitled "Common Approaches for Farmers Markets & Exempted Special Events, A Guide for Public Health Units" developed by the Association of Supervisors of Public Health Inspectors of Ontario (revised May 2012), that provides information on e.g., the legal background, key common principles and understandings as well as approaches to health hazard assessments and inspections [59]. Appendix I lists examples of non-potentially hazardous foods and Appendix II potentially hazardous foods.

Manitoba Although Manitoba does not have specific regulations for farmers markets, representatives indicated that they have plans to write farmers markets into the regulations. They do have farmers market guidelines in place that define farmers markets and that cover topics such as definitions, responsibilities, permits, food supplies, food protection, labelling, equipment and utensils [60].

Saskatchewan Although Saskatchewan does not have specific regulations for farmers markets, they do have the "Farmers Market Technical Guideline", July 15, 1998, that defines farmers markets [61]. In addition, the province indicated that their Food Safety Regulations are considered the reference regulation [62].

Alberta Part 3 of the food regulation sets out requirements for the market and vendors (stall holders), and there are a number of documents that can be found on the website of the Government of Alberta [63, 64]. These include the following documents.

1. The Alberta Approved Farmers Market Program Guidelines for the Farmers Markets, revised January 2015, gives an overview of the Alberta Approved farmers market program and contains a definitions section outlining the difference between "Farmers Markets" which is defined as an Alberta approved farmers market vs. a "Public Market", which means a farmers market operating within Alberta that is not approved. The document also highlights that there over 125 approved farmers markets in Alberta.
2. The Alberta Health Services Guideline for Public Market Managers and Vendors that provides information to the operators and vendors of public markets and flea markets.
3. The Farmers Market Home Study Course developed by the Environmental Public Health to help market managers and vendors set-up and operate an Alberta approved Farmers Market in a sanitary manner. There are 25 multiple questions and the passing grade is 80%. Upon successful completion, a Farmers Market Home Study Course certificate is awarded, which is valid for three years. The course is mandatory for Farmers Market managers, food vendors and staff preparing food for and selling at Farmers Market, who have not already completed the food safety training under section 31 of the Food regulations.
4. An Alberta Health Services farmers market fact sheet, which provides information on food products such as home-canned foods, high-risk foods, packaging of foods, labelling, food handling, food sampling, responsibilities of the market manager, handwashing and sink requirements. Additionally, there is a farmers market daily checklist for market managers and vendors.

British Columbia The Environmental Health Officers inspect farmers markets and operate under the *Public Health Act*, the *Food Safety Act* and the *Food Premises Regulation*. The British Columbia Centre for Disease Control and the Environmental Health Services have published the "Temporary Food Markets Guidelines for the Sale of Foods at Temporary Food Markets", revised May 2015, that covers among other things, definitions, e.g., of temporary food markets, explanations for preparation of lower-risk foods in the home, and recommendations for food handlers and food sampling [20]. Appendix I lists lower risk foods for example—apple sauce, bread and buns (no dairy or cheese filling), brownies, fudge, popcorn, etc. Appendix II lists higher-risk foods, for example—antipasto, cabbage rolls, cakes/pastries with whipped cream, cheese or synthetic fillings, pickled eggs and dairy products.

Appendix III is entitled, "Sale of Shell Eggs and Raw Foods of Animal Origin at Temporary Food Markets." The sale of higher risk foods to the public requires that the premises in which the food is processed, packaged, and sold to the public comply with the Food Premises Regulations. Hence, home prepared higher risk foods are not permitted to be sold to the public at temporary food markets unless prepared and sold in facilities that have been approved and, in some cases, issued a permit pursuant to the regulations.

Nunavut Farmers markets are not that popular, and therefore there are no specific regulations or guidelines. However, the Eating and Drinking Place Regulations that pertain to food safety exist under the *Public Health Act*.

Northwest Territories Currently, Northwest Territories does not have any specific regulations for farmers markets, however, regulations are in the development stages. The Northwest Territories indicated while there are not many farmers markets in the Territories, there is a guideline entitled "Operating Temporary Food Service Establishments Guideline" [65]. The guideline states that the sale of foods prepared at home is not allowed.

Yukon Although Yukon does not have any specific regulations or guidelines for farmers markets, they are currently guided by the *Agriculture Products Act* and the *Public Health Act* [66].

Summary

In general, farmers markets in most provinces and territories in Canada have fairly detailed requirements for food safety at the provincial/territorial level, and local associations are actively involved in promoting training and information for market managers and vendors. Although there have only been a few outbreaks in Canada linked to products purchased or consumed at farmers markets, more work needs to be done to promote safe food handling practices, temperature control and good agricultural practices on-farm. Comprehensive surveys to better understand the microbiological safety of foods at farmers markets would further highlight any additional food safety gaps that need to be addressed. The growth of farmers markets in Canada is expected to continue, as consumers continue to demand fresh, local food products that are free of pesticides, antibiotics, food additives, etc.

Acknowledgements We would like to acknowledge the excellent help of Barbara Kader-Farber with certain sections of this chapter, and Dr. Julie Jean for her assistance. The authors would also like to acknowledge the great help of the Federal Provincial Territorial Food Safety Committee members in providing information on farmers markets, as it related to their province or territory.

References

1. Experience Renewal Solutions (2009) Farmers market ontario impact study 2009 Report [cited 2017 Apr 25]. Available from: http://www.farmersmarketsontario.com/DocMgmt%5CResearch%5CFMO%20Research%20and%20Statistics%5CFMO%20Impact%20Study%20-%20Overview%20and%20Highlights.pdf
2. Connell DJ (2012) Economic and Social Benefits Assessment Provincial Report British Columbia, Canada [cited 2017 Apr 25]. Available from: http://www.bcfarmersmarket.org/sites/default/files/files/BCAFM%20Economic%20and%20Social%20Benefits-%20Final%20Report%202013(2).pdf
3. ACORN (Atlantic Canada Organic Regional Network) (2014) Exploring a farmers market network in New Brunswick, a summary of perspectives and recommendations [cited 2017 Apr 25]. Available from: http://www.acornorganic.org/media/projects/ACORNFarmersMarketNetworkReport2014.pdf
4. Farmers market Nova Scotia (2013) Nova Scotia farmers market economic impact study 2013 [cited 2017 Apr 25]. Available from: http://farmersmarketsnovascotia.ca/farmers-market-economic-impact-study-nova-scotia-2013/
5. Government of Alberta (2015) Alberta approved farmers market program guidelines [cited 2017 Apr 25]. Available from: http://www1.agric.gov.ab.ca/$Department/deptdocs.nsf/all/apa2577
6. Dungannon Consulting Services (2008) This little farmer went to market...an economic impact study of the member markets of the farmers markets association of Manitoba cooperative [cited 17 Dec 2015]. Available from: http://fmam.ca/wp-content/themes/fmam/FMAM_Website_Contents/FMAM%20Farmers%20Market%20Study%20(2008).pdf
7. AMPQ (Association des Marchés publics du Québec) (2014) Caractérisation des Marchés publics membres de l'AMPQ [cited 2017 Apr 25]. Available from: http://ampq.ca/pdf/ampq_etude_caracterisation_marches_publics_2014.pdf
8. Ellis A, Preston M, Borczyk A, Miller B, Stone P, Hatton B, Chagla A, Hockin J (1998) A community outbreak of *Salmonella* Berta associated with a soft cheese product. Epidemiol Infect 120:29–35
9. Honish L, Predy G, Hislop N, Chui L, Kowalewska-Grochowska K, Trottier L, Kreplin C, Zazulak I (2005) Can J Public Health 96(3):182–184
10. McIntyre L, Wilcott L, Naus M (2015) Listeriosis outbreaks in British Columbia, Canada, caused by soft ripened cheese contaminated from environmental sources. Biomed Res Intern 2015:1–12. doi:10.1155/2015/131623
11. PHAC (Public Health Agency of Canada) (2006) Laboratory surveillance data for enteric pathogens in Canada. Annual summary [cited 2017 Apr 25]. Available from: http://publications.gc.ca/site/eng/434756/publication.html
12. CFIA (Canadian Food Inspection Agency) (2014) Food recall warning – unpasteurized apple cider processed by rolling acres cider mill recalled due to *E. coli* O157:H7 [cited 2016 Jan 14]. Available from: http://www.inspection.gc.ca/about-the-cfia/newsroom/food-recall-warnings/complete-listing/2014-10-30/eng/1414720185030/1414720197088
13. PHO (Public Health Ontario) (2015) Public health science data request
14. MMWR (Morbidity and Mortality Weekly Report) (1999) Outbreaks of *Shigella sonnei* infection associated with eating fresh parsley – United States and Canada, July-August 1998. vol. 48(14):285–9. Available from: http://www.cdc.gov/mmwr/preview/mmwrhtml/00056895.htm
15. PAIFOD (Publically Available International Foodborne Outbreak database) (2015) Public Health Agency of Canada. Personal communication
16. Winnipeg Regional Health Authority (2010) Outbreak of verotoxigenic *E. coli* in the Winnipeg Health Region [cited 2017 Apr 25]. Available from: http://www.wrha.mb.ca/community/publichealth/cdc/files/VTECOutbreak_100928.pdf

17. BC-CDC (BC Centre for Disease Control) (2015) Botulism in British Columbia: the RISK of home-canned products [cited 2016 Jan 14]. Available from: http://www.bccdc.ca/resource-gallery/Documents/Educational%20Materials/EH/FPS/Food/HomepreparedfoodsbotulinumoutbreaksinBCandCanada_updatedJan2015.pdf
18. CFIA (Canadian Food Inspection Agency) (2014) Food recall warning – "Hausmacher" liver pâté recalled due to potential presence of dangerous bacteria. January 14, 2014 [cited 2016 Jan 14]. Available from: http://www.inspection.gc.ca/about-the-cfia/newsroom/food-recall-warnings/complete-listing/2014-01-14/eng/1389758388952/1389758426006
19. CFIA (Canadian Food Inspection Agency) (2011). Health hazard alert – certain Richardson's farm market brand pasteurized apple cider may contain *Salmonella* bacteria. November 18, 2011 [cited 2016 Jan 14]. Available from: http://www.inspection.gc.ca/about-the-cfia/newsroom/food-recall-warnings/complete-listing/2011-11-18/eng/1357653787251/1357653787267
20. BC-CDC (BC Centre for Disease Control) (2015) Temporary food markets: guideline for the sale of foods at temporary food markets [cited 2016 Jan 14]. Available from: http://www.bccdc.ca/resource-gallery/Documents/Guidelines%20and%20Forms/Guidelines%20and%20Manuals/EH/FPS/Food/GuidelinesSaleofFoodsatTemporaryFoodMarkets_MAY_01_2015a.pdf
21. New Brunswick Department of Health, Health Protection Branch (2016) The new Brunswick guidelines for food premises at public markets [cited 2017 Apr 25]. Available from: http://www2.gnb.ca/content/dam/gnb/Departments/h-s/pdf/en/HealthyEnvironments/Food/NBMarketGuidelines_E.pdf. Accessed 25 Apr 2017
22. Newfoundland Labrador (2011) Public market guidelines [cited 2017 Apr 25]. Available from: https://static1.squarespace.com/static/54d9128be4b0de7874ec9a82/t/562ae6c4e4b0411e298c1ca8/1445652164881/Public_Market-Guidelines_2012.pdf
23. Ministère de l'Agriculture, des Pêcheries et de l'Alimentation du Québec (MAPAQ) (2017). Available from: http://www.mapaq.gouv.qc.ca/fr/Publications/Marche_public.pdf
24. Food and Drug Regulations [cited 2017 Apr 25]. Available from: http://laws-lois.justice.gc.ca/eng/regulations/C.R.C.%2C_c._870/index.html
25. Food and Drugs Act [cited 2017 Apr 25]. Available from: http://laws-lois.justice.gc.ca/eng/acts/f-27/
26. Health Canada (2015) Voluntary guidance on improving the safety of soft and semi-soft cheese made from unpasteurized milk [cited 2015 Jan 14]. https://www.canada.ca/en/health-canada/services/food-nutrition/legislation-guidelines/guidance-documents/voluntary-guidance-improving-safety-soft-semi-soft-cheese-made-unpasteurized-milk-2015.html
27. Health Canada (2015) Joint United States Food and Drug Administration/Health Canada Quantitative assessment of the risk of listeriosis from soft-ripened cheese consumption in the United States and Canada [cited 2015 Nov 24]. Available from: https://www.canada.ca/en/health-canada/services/food-nutrition/food-safety/food-related-illnesses/food-specific-information/draft-joint-quantitative-assessment-risk-listeriosis-soft-ripened-cheese-consumption-united-states-canada-consultation.html
28. Teng D, Wilcock A, Aung M (2004) Cheese quality at farmers markets: observation of vendor practices and survey of consumer perceptions. Food Control 15:579–587. doi:10.1016/j.foodcont.2003.09.2005
29. Mihajlovic B, Dixon B, Couture H, Farber J (2013) Qualitative microbiological risk assessment of unpasteurized fruit juice and cider. Int Food Risk Anal J. Doi: 10.5772/57161
30. CCDR (Canada Communicable Disease Report) (1998) An outbreak of *Escherichia coli* O157:H7 infection associated with unpasteurized non-commercial, custom-pressed apple cider – Ontario, vol 25-13 July 1, 1999
31. Cody SH, Glynn MK, Farrar JA, Cairns KL, Griffin PM, Kobayashi J, Fyfe M, Hoffman R, King AS, Lewis JH, Swaminathan B, Bryant RG, Vugia DJ (1999) An outbreak of *Escherichia coli* O157:H7 infection from unpasteurized commercial apple juice. Annal Intern Med 130(3):292–209. doi:10.7326/0003-4819-130-3-199902020-00005
32. Steele BT, Murphy N, Arbus GS, Rance CP (1982) An outbreak of hemolytic uremic syndrome associated with ingestion of fresh apple juice. J Pediatr 101(6):963–965. doi:10.1016/S0022-3476(82)80021-8

33. CFIA (Canadian Food Inspection Agency) (1998) Code of practice for the production and distribution of unpasteurized apple and other fruit juice/cider in Canada [cited 2015 Dec 17]. Available from: http://www.inspection.gc.ca/food/processed-products/manuals/code-of-practice/eng/1340636187830/1340637184931
34. Health Canada (2000) Managing health risk associated with the consumption of unpasteurized fruit juice/cider products [cited 2017 Apr 25]. Available from: http://www.hc-sc.gc.ca/fn-an/legislation/pol/rev_unpast_juice_policy-rev_politique_jus_non_past_14-09-2000-eng.php
35. Scheinberg J, Cutter C (2014) Food safety at farmers markets: a reality check. Food Safety Magazine August/September 2014 [cited 2017 Apr 25]. Available from: http://www.foodsafetymagazine.com/magazine-archive1/augustseptember-2014/food-safety-at-farmers-markets-a-reality-check/
36. Chapman B (2014) Providing safe samples at farmers markets. Bargblog. March 14, 2014 [cited 2016 Jan 14]. Available from: http://barfblog.com/2014/03/providing-safe-samples-at-farmers-markets/
37. Kozak GK, MacDonald D, Landry L, Farber JM (2013) Foodborne outbreaks in Canada linked to produce: 2001 through 2009. J Food Prot 76(1):173–183
38. IFT (Institute of Food Technologists) (2001) Chapter II. Production practices as risk factors in microbial food safety of fresh and fresh-cut produce. In: Analysis & evaluation of preventive control measures for the control & reduction/elimination of microbial hazards on fresh & fresh-cut produce [cited 2015 Dec 3]. Available from: http://www.fda.gov/Food/FoodScienceResearch/SafePracticesforFoodProcesses/ucm090977.htm
39. FAO (2003) Code of hygienic practice for fresh fruits and vegetables. CAC/RCP-53 – 2003 [cited 2015 Jun 18]. Available from: http://www.fao.org/ag/agn/CDfruits_en/others/docs/alinorm03a.pdf
40. FAO (1999) Recommended international code of practice general principles of food hygiene. CAC/RCP 1-1969, Rev. 3 (1997), Amended 1999 [cited 2016 Jan 14]. Available from: http://www.fao.org/docrep/005/y1579e/y1579e02.htm
41. Rushworth G, Younie M (2006) Compliance assessment of agricultural practices in the Cloverdale area, British Columbia. September–December 2004 [cited 2015 Jun 18]. Available from: https://www.for.gov.bc.ca/hfd/library/documents/bib97170.pdf
42. Young I, Rajić A, Dooh L, Jones AQ, McEwen SA (2011) Use of good agricultural practices and attitudes toward on-farm food safety among niche-market producers in Ontario, Canada: a mixed methods study. Food Prot Trends 31(6):343–354
43. Harrison JA, Gaskin JW, Harrison MA, Cannon JL, Boyer RR, Zehnder GW (2013) Survey of food safety practices on small to medium-sized farms and in farmers markets. J Food Prot 76(11):1989–1993. doi:10.4315/0362-028X.JFP-13-158
44. Smathers SA, Chapman B, Phister T (2011) Evaluation of facilities and food safety practices in the North Carolina farmers market sector. IFT poster presentation [cited 2016 Jan 14]. Available from: http://chapmanfoodsafety.com/2011/06/01/ift-poster-presentation-evaluation-of-facilities-and-food-safety-practices-in-the-north-carolina-farmers-market-sector/
45. FAO/WHO (2008) Microbiological risk assessment series. Microbiological hazards in fresh fruits and vegetables. Meeting Report [cited 2015 Jun 18]. Available from: http://www.fao.org/fileadmin/templates/agns/pdf/jemra/FFV_2007_Final.pdf
46. Wood JL, Chen JC, Friesen E, Delaquis P, Allen KJ (2015) Microbiological survey of locally grown lettuce sold at farmers' markets in Vancouver, British Columbia. J Food Prot 78(1):203–208
47. Bohaychuk VM, Bradbury RW, Dimock R, Fehr M, Gensler GE, King RK, Rieve R, Romero BP (2009) A microbiological survey of selected Alberta-grown fresh produce from farmers' markets in Alberta, Canada. J Food Prot 72(2):415–420
48. Arthur L, Jones S, Fabri M, Odumeru J (2007) Microbial survey of selected Ontario-grown fresh fruits and vegetables. J Food Prot 70(12):2864–2867
49. Parks CE, Sanders GW (1992) Occurrence of thermotolerant campylobacters in fresh vegetables sold at farmers' outdoor markets and supermarkets. Can J Microbiol 38(4):313–316

50. CFIA (Canadian Food Inspection Agency) (2010) Food safety action plan report. 2009-2010 targeted surveys. Targeted survey investigating bacterial pathogens and generic *E. coli* in fresh leafy green vegetables [cited 2015 Jun 18]. Available from: http://www.inspection.gc.ca/food/chemical-residues-microbiology/microbiology/fresh-leafy-green-vegetables/eng/1397089237909/1397089239191

51. CFIA (Canadian Food Inspection Agency) (2012) Microbiology annual report 2011/12. National microbiological monitoring program [cited 2016 Jan 14]. Available from: http://inspection.gc.ca/food/chemical-residues-microbiology/microbiology/microbiology-annual-report-2011-12/eng/1410967997992/1410967998883

52. CFIA (Canadian Food Inspection Agency) (2013) Microbiology annual report 2012/13. National microbiological monitoring program [cited 2016 Jan 14]. Available from: http://www.inspection.gc.ca/food/chemical-residues-microbiology/microbiology/microbiology-annual-report-2012-13/eng/1415203899648/1415203900789

53. McIntyre L, Karden L, Shyng S, Allen K (2014) Survey of observed vendor food-handling practices at farmers' markets in British Columbia, Canada. Food Prot Trends 34(6):397–408

54. Behnke C, Seo S, Miller K (2012) Assessing food safety practices in farmer's markets. Food Prot Trends 32(5):232–239

55. Food Security Network for Newfoundland & Labrador (2011) Farmers market food safety: market organizer handbook a resource for farmers market organizers in Newfoundland & Labrador [cited 2015 Dec 14]. Available from: http://www.foodsecuritynews.com/Publications/Farmers_Market_Food_Safety_Organizer_Handbook.pdf

56. Nova Scotia Environmental Health and Food Safety Division (2016) Food safety guidelines for public markets [cited 2017 Apr 25]. Available from: http://novascotia.ca/agri/documents/food-safety/publicmarketguide.pdf

57. Prince Edward Island Department of Health and Wellness. Food premises regulations [cited 2015 Dec 14]. Available from: http://www.gov.pe.ca/health/foodregs

58. Teinovic N (2009) Guide de bonnes pratiques d'hygiène pour les marchés publics alimentaires [cited 2017 Apr 25]. Available from: http://www.cubiq.ribg.gouv.qc.ca/in/faces/details.xhtml?id=p%3A%3Ausmarcdef_0000995739&

59. ASPHIO (The Association of Supervisors of Public Health Inspectors of Ontario) (2012) Common approaches for farmers markets & exempted special events: a guide for public health units [cited 2016 Jan 14]. Available from: http://www.farmersmarketsontario.com/DocMgmt%5CFood%20Safety%5CHealth%20Unit%20Guidelines%5CASPHIO-CommonApproachesGuidelinesRevisedMay2012.pdf

60. Manitoba Agriculture, Food and Rural Initiatives (2009) Farmers market guidelines [cited 2017 Apr 25]. Available from: https://www.gov.mb.ca/health/publichealth/environmental-health/protection/docs/farmers_market.pdf

61. Saskatchewan, Government of (1998). The farmers market technical guideline, July 15, 1998

62. Saskatchewan, Government of (2009) The food safety regulations [cited 2017 Apr 25]. Available from: http://www.qp.gov.sk.ca/documents/English/Regulations/Regulations/P37-1R12.pdf

63. Province of Alberta (2006) Public health act food regulations [cited 2015 Dec 18]. Available from: http://www.qp.alberta.ca/1266.cfm?page=2006_031.cfm&leg_type=Regs&isbncln=9780779785742

64. Alberta Approved Farmers Market Program (2017) [cited 2017 Apr 25]. Available from: www1.agri.gov.ab.ca/$Department/deptdocs.nsf/all/bus15891

65. Northwest Territories – Health and Social Services (2011) Guidelines for operating temporary food service establishments [cited 2017 May 30]. Available from: https://www.justice.gov.nt.ca/en/files/legislation/public-health/public-health.r8.pdf

66. Yukon Government. Legislation – acts and regulations. (Not dated) [cited 2017 May 30]. Available from: http://www.gov.yk.ca/legislation/legislation/page_a.html

Chapter 8
An Overview of Farmers Markets in Australia

Bruce Nelan, Edward Jansson, and Lisa Szabo

Abstract The emergence of modern-day farmers markets in the United States and England resulted in interest in Australia. There were reported to be 70 farmers markets Australia-wide in 2004 while the Australian Farmers Market Association Directory listed 180 markets in 2015. Australian studies report that farmers markets offer many and varied benefits to a wide variety of stakeholders. The regulatory oversight of farmers markets is largely undertaken by the local governments which exercise control over the event and the individual stallholders (vendors). State authorities exercise control over the processing of high-risk foods where a licence is required for the activity. Regulatory interventions are risk based with vendors selling foods that require temperature control for safety receiving the most attention. An examination of the New South Wales register of penalty notices found that very few farmers market vendors have been penalised over a seven year period and those that have been were foodservice businesses (selling cooked or pre-prepared home meals or market snacks) trading at the market.

Keywords Farmers market • Regulatory oversight • Food Standards Australia New Zealand • Horticulture • Eggs • Poultry • Dairy products • Fish • Meat • Case study • Sydney • Mornington Peninsula

"A farmers market is a predominantly fresh food market that operates regularly within a community, at a focal public location that provides a suitable environment for farmers and food producers to sell farm-origin and associated value-added specialty foods and plant products directly to customers" [1]. The emergence of these modern-day farmers markets in the United States and England resulted in interest in

B. Nelan • E. Jansson • L. Szabo (✉)
NSW Food Authority, P.O. Box 6682, Silverwater NSW 1811,
Avenue of the Americas, Newington, NSW 2127, Australia
e-mail: Bruce.Nelan@foodauthority.nsw.gov.au; Lisa.Szabo@foodauthority.nsw.gov.au

© Springer International Publishing AG 2017   103
J.A. Harrison (ed.), *Food Safety for Farmers Markets: A Guide to Enhancing
Safety of Local Foods*, Food Microbiology and Food Safety,
DOI 10.1007/978-3-319-66689-1_8

Australia, which was originally fostered through a series of workshops commencing in 1999. The Victorian Farmers Markets Association credits Jane Adams with spearheading Australian farmers markets after studying the movement in the United States [2]. Market numbers and their economic importance increased to levels that warranted reports by the Australian Government's Rural Industries Research and Development Corporation (RIRDC) in 2005 and 2014 and by a State of Victoria parliamentary committee in 2010 [3–5].

The number of farmers markets was reported to be 70 in 2004 [3]. In 2015 there were 180 in the Australian Farmers Markets Association directory [6]. Estimates of the economic impact of farmers markets were summarized by Jane Adams in 2012: in 2004 the national turnover was $A40 million (40 million in Australian dollars); in 2010 in the State of Victoria with 97 markets the economic impact was $A227 million with half being market turnover and the balance being multiplier effect; estimates of the average amount spent by visitors at Victorian markets were $A29.50 to $A33.50 for fruit and vegetables with an overall average expenditure being $A70 per shopper [7]. McKinna estimated in 2011 that combined retail/reseller markets and farmers markets amounted to 7% of fresh food sales, and the share was growing [8]. He added that markets are an important marketing channel for suppliers that are too small to service major supermarkets.

Markets branded as farmers markets can take different forms depending on the rules they impose on sellers and whether resellers are permitted to participate. The Australian Farmers Markets Association recommends that markets should be limited to businesses that produce food and specialty processors that primarily use local, seasonal and regional ingredients [1]. It discourages vendors selling crafts and those reselling products. The Victorian Farmers Markets Association has developed an accreditation system for vendors so that the public can have confidence in the authenticity of producers [9]. Further they also accredit markets. For farmers markets in metropolitan areas to gain accreditation, at least 90% of their vendors must be accredited; for regional farmers markets, 75% of the vendors must be accredited. Although there is a clear intention among branded farmers markets to maintain authenticity, some markets admit resellers to broaden the range of foods offered. Some market operators are opposed to selling produce from outside the region, and some do not allow farmers to sell on behalf of neighboring farmers. Achieving a balance has proven to be difficult, and in 2015 a group of farmers markets in the state of Queensland banned the sale of imported garlic in response to complaints in social media [10].

A committee of the Parliament of Victoria considered a range of issues that affected farmers markets [5]. The "authentic farmers market" was the primary concern. Supply of vendors was reported to be an issue in regional markets. Evidence given to the Committee said that a minimum of 30 vendors are required for the market to be viable and to offer a satisfying experience for patrons. Many Victorian markets outside of the capital city had problems maintaining those numbers. Some farmers by-pass the local community market and instead attend larger city markets with more customers and higher prices for their products. Some communities lack sufficient numbers of farmers to adequately supply the market. Very large farms are likely to be contracted to major supermarkets, and some smaller farms are unable to commit time to markets. Establishing new markets is difficult, and 25–30% of new farmers markets close, mostly in the first year. Other markets change their focus and move away from the farmers market model to allow the sale of crafts, clothes, bric-a-brac or other non-food items.

Land use planning hinders the development of farmers markets. Planning permits are expensive. Markets are often located at sports grounds, remote from the main shopping precinct, and that splits the audience. This may disadvantage the market and the retail center making it more difficult to attract customers. The requirements for food permits which are applicable only to a single local government area are problems for those that trade in several different local government areas each month. The committee made a number of recommendations to address these issues.

Australian studies suggest that farmers markets offer many and varied benefits to a wide variety of stakeholders [3–5]. Farmers benefit from receiving higher prices for a portion of their produce; having direct contact with consumers and hearing what they want; and from being able to try new products and farming practices in a low-risk environment. Food businesses have the opportunity to design, and test food products directly with the consumer which can lead to improvements in the products. The 2014 RIDRC report on a survey of farmers selling at the markets found that 64% reported they had made changes to their produce, management approach, presentation and or branding as a result of participation in the farmers market [4]. Consumers benefit from access to fresh, local and seasonal food. Communities benefit from tourism and money being spent locally, and local retailers share in the benefits. Farmers markets are often said to be an incubator for local businesses, and they can provide a positive environment to learn and improve the business through connecting with consumers and other market participants. The social interaction they offer benefits individuals and communities.

Highlights from a survey undertaken in 2010 by the Victorian Farmers Markets Association that profiled market users were reported by Adams [7]. Customers were mainly female; their average age was 50; most were employed, mainly in white collar jobs; income averaged $A80,000 per annum; they visited an average of nine times per year; and more than half had attended the market for more than 3 years. Their motivation for attending the market included access to fresh quality produce (89%); social factors (31%); and desire to improve the environment and support local farmers (30%). Increased consumption of fruit and vegetables as a result of visiting farmers markets was reported by 58% of shoppers. About one third of shoppers also made purchases at local non-market businesses. The picture that emerges from this research is that those who attend Australian farmers markets tend to be mature reasonably affluent people. This view is supported by evidence given to the Victorian Parliamentary committee: '…You cannot just go into a new housing estate (government-provided, public housing) and expect people to go to a farmers market; they will not. Demographically, they will not because they are usually on a budget; they are usually watching their pennies. We all know that farmers markets are not based on price; they are based on quality and energy and where the food comes from' [5].

Coster and Kennon report that nearly all the markets sold fresh vegetables, fruit, value-added fruit and jam, baked goods, poultry and eggs [3]. Over three-quarters of the markets sold honey, oils, prepared food and drinks, non-alcoholic beverages and dairy products. Over half the markets sold fish and other seafood, nuts, alcoholic beverages, farm-raised meats, spices, confectionery and small goods.

The Australian Food Regulation System

Food regulations are jointly developed by New Zealand and Australian agencies. Food regulation in Australia involves all three levels of government, the Commonwealth, states and territories and local government [11]. Food Standards Australia New Zealand (FSANZ), is the joint Australia and New Zealand standards setting body. Standards are published in the Australia New Zealand Food Standards Code (the Code). Individual states and territories have laws to adopt the Code. Enforcement of food law is shared by state agencies and local government.

Food Standards Australia New Zealand (FSANZ) is an independent statutory authority responsible for developing all domestic food standards based on scientific/technical criteria. The Food Standards Code includes standards for food additives, food safety, labeling and foods that need pre-approval such as GM foods [12]. Standards are accessible online at www.foodstandards.gov.au. The Code is divided into four chapters:

Chapter 1—Introduction and standards that apply to all foods.
Chapter 2—Food product standards: compositional requirements for specific foods.
Chapter 3—Food safety standards: Australia-only requirements for food businesses and food handlers, including foodservice to vulnerable populations.
Chapter 4—Primary production standards: Australia-only requirements for primary production and processing of agricultural commodities.

Szabo et al. have provided a history of changes to food regulation in Australia [13]. They note that the Food Standards Code is not overly prescriptive, and it strives for performance and outcome-based regulation. Areas requiring pre-market approval are specifically identified as are the procedures for seeking amendment to the Food Standards Code. For areas outside of these, the onus is on the industry to control food safety hazards and produce food that is safe and suitable and on governments to regulate on a risk basis to ensure that the industry has mechanisms in place to produce safe and suitable food. Taken together, this represents a co-regulatory approach based on a partnership between consumers, industry and government to achieve food safety outcomes and at the same time enable the industry to expand, innovate and evolve. The authors note that while outcome-based regulations promise to bring economic reform, their application in industries and by regulators accustomed to prescriptive regulation poses some implementation challenges.

Table 8.1 Products sold at Australian farmers markets and by market vendors

Product	Markets selling the product [3]	Vendors selling the product [4]
Fruit	90–100%	19%
Vegetables	90–100%	12%
Value-added fruit and vegetables	90–100%	7%
Baked goods	90–100%	6%
Poultry and eggs	90–100%	4%
Prepared food and drinks	80–90%	5%
Oils	80–90%	4%
Honey (and related)	80–90%	4%
Non-alcoholic beverages	80–90%	4%
Dairy products	70–80%	3%
Fish and other seafood	60–70%	-
Alcoholic beverages	60–70%	3%
Nuts	60–70%	4%
Farm-raised meats	50–60%	Meat 6%
Processed meats	50–60%	
Game meats	40–50%	
Spices (and herbs)	50–60%	5%
Confectionery	50–60%	2%
Condiments	–	3%
Pulses and grains	40–50%	–

For example, the removal of some microbiological standards from the Code initially made it difficult for industry to know when their product was safe and difficult for regulators to prove that a product was unsafe.

Regulation of Products Sold at Farmers Markets

The reports commissioned by RIRDC contain information on products sold by farmers markets and vendors [3, 4]. Data concerning products sold are presented in Table 8.1.

Fruit and Vegetables

Other than requirements for seed sprouts, there are currently no food safety requirements in the Australia New Zealand Food Standards Code applying specifically to the primary production of horticultural produce. Proposal P1015 was prepared by FSANZ in 2012 to examine the hazards of horticultural produce, existing risk management measures and other possible measures that could be introduced to the primary production and processing of fresh horticultural produce. The Proposal was abandoned in February 2014, in favor of a strategy for maximizing food safety in horticultural produce using existing regulatory and non-regulatory systems [14]. Horticultural produce has a generally good record of food safety in Australia, but outbreaks of foodborne illness attributable to produce are reported intermittently. FSANZ noted that fresh horticultural commodities involved in outbreaks were those intended to be eaten uncooked without any steps to eliminate pathogens before consumption [14]. Two general commodity categories were identified from the outbreak data: soft fruit (melons, papaya, mango, tomatoes and berries) and vegetables, including leafy greens (lettuce, spinach), herbs (coriander, basil and Thai basil), green onions, baby corn, sugar peas, carrots and chili peppers. From the available data, the use of poor quality water for pre- and post-harvest activities emerged as the most common cause of produce contamination. Microbiological data available from Australian surveys suggest a low level of contamination of fruits and vegetables in the Australian supply chain, although infrequent contamination with pathogenic microorganisms can occur. The available evidence provides a high degree of confidence that Australians have access to safe fresh produce.

An estimated 70–80% of horticultural produce is grown under a food safety scheme or plan that contains measures to control identified hazards [14]. A number of the schemes in use have been mandated by major supermarkets, and it is likely much of the produce that is grown under a food safety scheme or plan will be sold through supermarkets and the large wholesale markets. Some producers who sell through farmers markets also supply wholesale markets (29% of those surveyed) and major retailers (18%), but 4% of vendors only supply farmers markets and 20% sell 76–99% of their produce through that channel [4]. At least some, and possibly a moderate proportion, of the produce sold through farmers markets will not be produced under a food safety plan and will not benefit from the food safety surveillance systems in large wholesale markets.

Value-Added Fruit and Vegetables

The surveys did not identify the products categorized as value-added fruit and vegetables. The following products have been identified at markets in the Sydney region: preserves; pickles; jams; olives; roast garlic in oil; chutneys; relish; fruit juices; tapenades; dips and salsas. The Food Standards Code contains few specific requirements for these types of products. There are compositional requirements for jams and conserves. There is also a food safety related compositional requirement for fruit and vegetables in brine, oil, vinegar or water, other than commercially canned fruit and vegetables, which requires they must not have a pH greater than 4.6. Chapter 1 requirements including those for labeling and Chapter 4 requirements for food safety in the Food Standards Code are applicable.

Jams, conserves and traditional acid-preserved products have a long history of safe consumption. The same cannot be said about garlic in oil which has been associated with human botulism. Garlic in oil and olives stuffed with garlic have been identified in Sydney region markets. These are not traditional products in Australia, and they are unlikely to be prepared using a scientifically tested and safe recipe or even a well-proven recipe handed down from generation to generation. If basic technical requirements are overlooked, then product spoilage or food poisoning is possible. Local guidance on these products is available [15].

Eggs

Australian laying flocks remain free of transovarian/transovarial *Salmonella* Enteritidis, but contaminated eggs and egg products are still suspected to be the cause of an unacceptably high number of foodborne illness outbreaks in Australia [16, 17]. *Salmonella* Typhimurium has been identified as the dominant serovar responsible for outbreaks of illness associated with egg consumption [18] with the use of dirty and/or cracked eggs in uncooked or undercooked foods being a significant risk factor. Barfblog maintains a table of Australian foodborne illness outbreaks associated with eggs [19]. The table provides brief information on 232 outbreaks from 1991 to March 2015. The New South Wales Food Authority has published three case studies that illustrate the hazards of raw egg usage [20]. The first of these describes an outbreak traced to a hamburger restaurant that was using aioli prepared using raw eggs. In this incident, the source of eggs was not a farmers market, but it was another direct marketing channel. The restaurant was sourcing eggs from a local hobby farm rather than a dedicated egg supplier. The hobby farm did not have any system for quality control such as candling or crack detection, and eggs were not safely washed prior to sale. Eggs were also placed into re-used cartons which increased the potential for cross-contamination of *Salmonella* to the outside of shells.

The Food Standards Code has a food product standard for eggs that prohibits the sale of cracked or dirty eggs or unprocessed egg pulp for catering purposes or retail sale. It also requires eggs for retail sale or for catering purposes to be individually marked with the producers' or the processors' unique identification. The Primary Production and Processing Standard for Egg and Egg Product sets a number of food safety requirements that also apply to the direct sale of eggs to the public by an egg producer. The Standard includes requirements of egg producers for: risk-based food safety planning; control of inputs such as food, water and veterinary medicines; food safety and food hygiene skills and knowledge; bird health and egg traceability. Applicable requirements from that list are also imposed on egg processors. Some Australian states and territories exempt very small businesses from all or most of the requirements. In New South Wales (NSW), businesses that supply less than 20 dozen eggs in any week have been exempted from most requirements. This is likely to cover many egg sellers at NSW farmers markets.

Poultry

In Australia, raw poultry meat purchased by the consumer is very likely to be contaminated with *Campylobacter* (90%) and to a lesser extent, *Salmonella* (43%, although over 2/3 of isolates are the rarely pathogenic *Salmonella* Sofia serotype). The higher the prevalence and concentration of these two bacteria on raw poultry, the greater the likelihood these pathogens could be present at the point of consumption and therefore a greater likelihood of illness occurring [21].

Campylobacteriosis and salmonellosis are the gastrointestinal diseases most frequently reported to the National Notifiable Diseases Surveillance System [22]. Almost 1/3rd of cases of campylobacteriosis can be attributed to contaminated poultry [23]. Similar data are not available for salmonellosis, but a portion of the large number of cases in Australia could reasonably be expected to come from contaminated chicken.

Figure 8.1 provides data on Australian notifications of campylobacteriosis. Two main documents support the regulation of poultry meat processing in Australia. They are Food Standard 4.2.2—*Primary Production and Processing Standard for Poultry Meat* and *Australian Standard for the Hygienic Production and Transportation of Meat and Meat Products for Human Consumption* [24]. All states and territories exercise considerable regulatory control over poultry processors. Controls may extend to requiring businesses to be licensed, and small processors are not exempted.

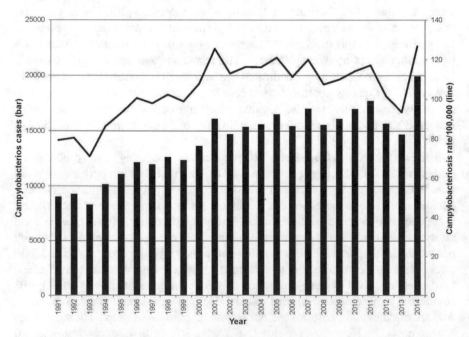

Fig. 8.1 Australian notifications of campylobacteriosis: cases and annual rate/100,000 from 1991 to 2014 (net of the State of New South Wales where campylobacteriosis is not notifiable). Courtesy of Lisa Szabo

Dairy Products

The Australian dairy industry was heavily regulated until the year 2000. Statutory marketing authorities were established in each state to administer the regulation of the market's milk sector [25]. Year round supply of fresh milk to the large cities and assurance of safety and quality were key roles. The Australian dairy industry was an early adopter of HACCP-based food safety systems. Market-based systems now service consumer needs for the supply of quality milk and dairy products. State regulatory agencies regulate milk and dairy product safety. Australian dairy products have a good record of safety. In a risk profile of Australian dairy products, FSANZ found: Australian dairy products have an excellent reputation for food safety, and this is supported by the lack of evidence attributing foodborne illness to dairy products; and there are extensive regulatory and non-regulatory measures in place along the dairy industry primary production chain resulting in minimal public health and safety concerns regarding the use or presence of chemicals in dairy products [26].

The Food Standards Code includes food product standards for milk, cream, fermented milk products, cheese, butter, ice cream, dried milks, evaporated milks and condensed milks. It also includes Standard 4.2.4—*Primary Production and Processing Standard for Dairy Products*. The most recent version of Standard 4.2.4, includes provisions for the manufacture of certain raw milk cheeses. Information from the dairy industry suggests that raw milk cheeses will be embraced mainly by artisanal cheesemakers, many of whom use direct marketing channels including farmers markets. Standard 4.2.4 only approves cheeses where there is no net growth in pathogenic bacteria following the completion of maturation and where the finished cheese does not support the growth of pathogenic bacteria, specifically *Listeria monocytogenes*. Cheesemakers are required to demonstrate to the state regulatory agency that their cheeses meet these requirements prior to the cheeses being placed on sale. Keeping track of new businesses and new raw milk cheeses is expected to present a challenge for regulators.

Fish and Other Seafood

FSANZ and the New South Wales Food Authority have reported on the food safety risks of Australian seafood [27–29]. Ciguatera is a recurring problem in both commercially and recreationally harvested fish from mainly tropical and more northern sub-tropical waters. Scombroid poisoning is also encountered, as is escolar/rudderfish keriorrhoea. Prior to the establishment of an internationally accepted shellfish quality assurance program, New South Wales had recurring problems with viral contamination of shellfish harvested from heavily populated estuaries. Only infrequent problems with norovirus and algal biotoxin have been reported from Australian states since the introduction of shellfish quality assurance programs.

Seafood primary production and processing is regulated by Standard 4.2.1. Other food safety issues are covered by Chapter 3 Standards. The food product standards for fish and fish products do not address food safety other than requiring that formed or joined fish in the semblance of a cut or filet of fish include cooking instructions that ensure the microbiological safety of the product.

Seafood suppliers in some Sydney area farmers markets offer fresh and value-added fish, crustaceans and molluscs. The suppliers identified in Sydney were mainly seafood processors. In regional areas, producers and integrated producer-processors were better represented.

Meat and Meat Products

FSANZ prepared a comprehensive set of documents addressing food safety risks with meat and meat products [30]. The key findings were that: the major meat species (cattle, sheep, goats and pigs) have a low microbial load and a generally low prevalence of pathogens; foodborne illness from the consumption of meat is quite low; and extensive regulatory and non-regulatory measures are in place resulting in minimal public health and safety concerns regarding the presence of chemicals in meat. FSANZ found limited data on minor species and wild game but concluded that production and processing risk factors were not substantially different from those of the major meat species.

The Food Standards Code includes a food product standard that establishes information requirements for raw meat joined or formed into the semblance of a cut of meat and for fermented meat products. It also requires that bovine meat must be derived from cattle free of bovine spongiform encephalopathy. The primary production and processing standard sets out requirements for producers of ready-to-eat meat and additional requirements for uncooked comminuted fermented meat. All states and territories exercise considerable regulatory control over meat processors. Controls may extend to requiring businesses to be licensed, and small processors are not exempted.

A wide range of meat and meat products have been identified in Sydney area markets. These include cuts of beef, lamb and pork; fresh sausage; smoked meats; bacon; salami; and paté. Vendors include specialty butchers with a reputation for quality products.

Prepared Foods and Foodservice at Markets

Commercially manufactured foods are infrequently associated with foodborne illness outbreaks in Australia, however, the prepared foods on sale at farmers markets would be better described as foodservice products for home consumption. Foodservice accounts for most of the reported outbreaks of foodborne illness in Australia [31]. During visits to markets, it is apparent that many people have a meal or snack and a beverage during the visit. Queues for coffee are seemingly longer than at any other stall in the market. The service of foods at markets is a focus of attention for local government food inspection staff. The foods are regulated under Chapter 3 food safety standards in the Food Standards Code.

Regulatory Oversight of Farmers Markets

The regulatory oversight of farmers markets is largely undertaken by local government who exercise food safety control over the event and the individual vendors. State authorities exercise control over the processing of high-risk foods that require a license for the activity. Regulatory interventions are risk-based, with vendors selling foods that require temperature control for safety receiving the most attention.

Case Study 1: State of New South Wales: Inner-City Markets

Sydney is Australia's oldest and most populous city. The City of Sydney has established requirements for temporary food stalls that apply to stalls at farmers markets [32, 33]. Events, including markets, must be registered. Intending vendors must apply for approval to sell food or drink at a public event. Applications may be for the sale of high-risk foods, potentially hazardous food requiring temperature control for safety, or low-risk foods which have a lower fee and a longer approval period. The fee covers the cost of food safety inspections. Applicants must provide evidence of a current satisfactory food safety inspection for all off-site food preparation areas from the city, another local government area or the state regulator. Stalls providing foods that are ready-to-eat, potentially hazardous and not in the supplier's original packaging must appoint a food safety supervisor in line with state law [34].

Specific requirements have been established for the structure of the stall, cooking equipment, food display and protection, rubbish disposal, washing facilities, food temperature control and behaviors of food handlers. The requirements are mostly traceable to the Food Standards Code and food safety is the key message, but environmental protection and public safety are considered. Inspection frequency is based on risk. High-risk stalls are inspected more frequently, and markets with a greater number of high-risk stalls are visited more frequently.

Case Study 2: State of Victoria: Outer-Urban/Regional Markets

Mornington Peninsula Shire is a mixture of urban areas, resort towns, tourist development and rural land. It is located just over an hour's drive from Melbourne, the state's capital. The 'Peninsula' is a popular for destination for city dwellers and hosts many markets. Food stalls at markets are classified as temporary or mobile food businesses and are regulated accordingly [35]. Victoria has implemented a statewide online system for businesses and community groups to register their temporary and mobile food premises (stalls, trucks, vans or carts) called Streatrader [36]. It allows businesses and community groups to apply for a Food Act registration and advise authorities of where they intend to trade. Importantly, one council becomes the registering council, and Streatrader notifies relevant councils when trade will occur within their local government area. Streatrader was developed to avoid the cost and inconvenience of registering with multiple councils.

Businesses are registered for 12 months. The requirements imposed on businesses are risk-based with higher standards for Class 2 food premises that handle potentially hazardous food which needs correct temperature control during food handling. The Food Standards Code is applicable to all businesses, and Class 2 businesses are required to have a risk-based food safety plan and a food safety supervisor. Requirements for lower risk, Class 3 food premises are less demanding [37]. Mornington Peninsula Shire Council staff inspect each market twice per year and each Class 2 food premises once per year. However, businesses that trade in several local government areas could be inspected on multiple occasions each year. Inspection reports are shared between councils using Streatrader. The sharing of inspection reports provides an additional incentive for businesses to address issues and score well during inspections. Multiple bad inspection reports could result in the registering council withdrawing registration.

Regulatory Issues Identified at Markets

New South Wales operates a Name and Shame website for food businesses that receive a Penalty Notice (PN). The database for the website contains over 14,000 entries spanning seven years of operation. Only 93 PNs were found where premises were identified as "temporary" or "mobile". Only 14 PNs were found were premises were identified as "market" and "stall," and few of the these were issued to vendors at farmers markets. All the PNs were related to foodservice businesses, and half of the 14 PNs were related to failure to provide adequate handwashing facilities. Compliance with the Food Standards Code requirement for handwashing in a temporary premises does present some difficulties. Mornington Peninsula Shire Council had provided specific information on minimum requirements for a handwashing station in their local government area [38].

Other problems resulting in PNs included poor labeling, failure to protect food on display from contamination and inadequate temperature control. Given the complexity of Australian food labeling regulations, it is surprising that labeling breaches were not recorded more frequently. Labels on value-added products purchased from Sydney area farmers markets were found to have addressed the more complex requirements such as ingredient declarations, nutrient information panels and allergen statements reasonably well. However, home-based businesses appeared to be reluctant to include their home address on labels, and contact details were often web addresses or social media posts. Lot or batch coding was sometimes omitted. Food Standards Australia New Zealand has developed guidance material on food safety and labeling requirements for the Australian Farmers Markets Association [39]. The handbook is available at   https://farmersmarkets.org.au/wp-content/uploads/AFMA-Farmers-Market-Food-Safety-Guide-Aug-15.pdf. The role of farmers markets as business incubators has been mentioned previously. Food regulation interferes with this role in the case of dairy products, meat and other products where manufacture is a licensed activity. In these cases, the business cannot simply 'find its way' because risk-based food safety plans are mandatory. Processing businesses have been set up to dehydrate meat, prepare beef jerky or smoke fish only to find out they require an approved facility, audited food safety plan and a license prior to selling product. The degree of sophistication required is a barrier to entry to the market. Where unlicensed manufacturers have been identified, most have been given the opportunity to become licensed.

Summary

Although the Food Standards Code sets many requirements for the processing and handling of various food products, regulatory oversight of farmers markets in Australia is largely undertaken by local governments. Many vendors selling in direct markets may be exempt from legislation. More attention is paid to vendors selling high risk foods. Failure to perform adequate handwashing and temperature control of foods are problems observed in farmers markets. This raises potential concerns over contamination, cross-contamination, growth of pathogens and risk of foodborne illnesses.

References

1. Australian Farmers Market Association (not dated) What is a farmers market? [cited 2017 June 15]. Available from: http://farmersmarkets.org.au/definition
2. Victorian Farmers Market Association (Submission to Parliamentary Committee) (2010) Inquiry into farmers markets. Melbourne Government printer for the State of Victoria [cited 2017 June 15]. Available from: http://www.parliament.vic.gov.au/images/stories/committees/osisdv/Farmers_Markets/Submissions/OSISDC_FarmersMarkets_Sub7_VictorianFarmersMarketAssociation_14.05.10.pdf
3. Coster M, Kennon N (2005) "New generation" farmers markets in rural communities. RIRDC Publication No. 05/109 [cited 2017 June 15]. Available from: https://rirdc.infoservices.com.au/downloads/05-109
4. Woodburn V. Understanding the characteristics of Australian farmers markets. RIRDC Publication No. 14/0402014 [cited 2017 June 15]. Available from: https://rirdc.infoservices.com.au/items/14-040
5. Parliament of Victoria (2010) Outer suburban/interface and development committee: inquiry into farmers markets Melbourne Government Printer for the State of Victoria [cited 2017 June 15].

Available from: http://www.parliament.vic.gov.au/images/stories/committees/osisdv/Farmers_
Markets/OSISDC_FarmersMarketsWEB15.10.10.pdf

6. Australian Farmers Markets Directory: Australian Farmers Markets Association (not dated)
   [cited 2016 Dec 12]. Available from: http://farmersmarkets.org.au/markets

7. Adams J (2012) Grow up: fresh direct marketing channels. Presentation at Australian Bureau
   of Agricultural and Resource Economics and Sciences outlook conference [cited 2015 May
   26]. Available from: http://www.daff.gov.au/SiteCollectionDocuments/abares/outlook/2012/
   conference-presentations/Jane-Adams-Horticulture.pdf

8. McKinna D (2011) Are supermarkets milking producers? Presentation at Australian Bureau
   of Agricultural and Resource Economics and Sciences outlook conference [cited 2015 May
   26]. Available from: http://www.daff.gov.au/SiteCollectionDocuments/abares/outlook/2011/
   David_McKinna_-_Food_Industry.pdf

9. Victorian Farmers Markets Association (2014) VFMA accreditation [cited 2016 Dec 12].
   Available from: http://www.vicfarmersmarkets.org.au/content/vfma-accreditation

10. Hinchliffe J (2015) Chinese garlic moves local produce debate from supermarkets to farmers
    markets in Brisbane (Transcript of radio news item). ABC Radio 612 Brisbane [cited 2016
    Dec 12]. Available from: http://www.abc.net.au/news/2015-05-18/chinese-garlic-moves-
    local-produce-debate-to-brisbane-markets/6477058

11. Australian Government Department of Health (2014). The food regulation system [cited 2016
    Dec 12]. Available from: http://www.health.gov.au/internet/main/publishing.nsf/Content/
    foodsecretariat-system1.htm

12. Food Standards Australia New Zealand. Food standards code – a quick guide [cited 2016 Dec
    12]. Available from: http://www.foodstandards.gov.au/consumer/generalissues/codeguide/Pages/
    default.aspx

13. Szabo EA, Porter WR, Sahlin CL (2008) Outcome based regulations and innovative food pro-
    cesses: an Australian perspective. Innovative Food Sci Emerg Technol 9(2):249–254

14. Food Standards Australia New Zealand (2014) Proposal P1015 – primary production & pro-
    cessing standard for horticulture abandonment: [cited 2016 Dec 12]. Available from: http://
    www.foodstandards.gov.au/code/proposals/Pages/proposalp1015primary5412.aspx

15. CSIRO (2015) Vegetable preservation [cited 2016 Dec 12]. Available from: http://www.csiro.
    au/en/Research/Health/Food-safety/Vegetable-preservation

16. New South Wales Food Authority (2013) Egg food safety scheme; periodic review of the
    risk assessment [cited 2016 Dec 12]. Available from: http://www.foodauthority.nsw.gov.au/_
    Documents/scienceandtechnical/egg_food_safety_scheme.pdf

17. Food Standards Australia New Zealand (2011) Proposal P301 primary production & process-
    ing standard for eggs and egg products; final assessment report [cited 2016 Dec 12]. Available
    from: http://www.foodstandards.gov.au/code/proposals/documents/P301%20Eggs%20_%20
    Egg%20Products%20PPPS%20FAR%20FINAL%20AMENDED%2024061.pdf

18. Food Standards Australia New Zealand (2009) Proposal P301 primary production & pro-
    cessing standard for eggs & egg products; risk assessment of eggs and egg products [cited
    2016 Dec 12]. Available from: http://www.foodstandards.gov.au/code/proposals/documents/
    P301%20Eggs%20PPPS%20DAR%20SD1%20Risk%20Assessment.pdf

19. Barfblog (2015) Foodborne illness outbreaks associated with egg consumption (Australia)
    [cited 2016 Dec 12]. Available from: http://barfblog.com/wp-content/uploads/2015/03/raw-
    egg-related-outbreaks-australia-3-12-15-2.pdf

20. New South Wales Food Authority (2012) Foodborne illness (FBI) case studies [cited 2015 May 26].
    Available from: http://www.foodauthority.nsw.gov.au/science/foodborne-illness-case-studies

21. Food Standards Australia New Zealand (2010) Proposal P282 primary production & process-
    ing standard for poultry meat; final assessment report [cited 2016 Dec 12]. Available from:
    http://www.foodstandards.gov.au/code/proposals/documents/P282%20Poultry%20PPPS%20
    FAR%20FINAL.pdf

22. Australian Goverment Department of Health (2015) National notifiable diseases surveillance sys-
    tem [cited 2016 Dec 12]. Available from: http://www9.health.gov.au/cda/source/cda-index.cfm

23. Stafford R (2013) Foodborne campylobacteriosis in Australia. Microbiol Aust 34(2):98–101
24. Australian Standard for the Hygienic Production and Transportation of Meat and Meat Products for Human Consumption (2007) AS 4696:2007: CSIRO Publishing/Food Regulation Standing Committee (FRSC) [cited 2016 Dec 12]. Available from: http://www.publish.csiro.au/Books/download.cfm?ID=5553
25. Edwards G (2003) The story of deregulation in the dairy industry. Aust J Agric Resour Econ 47(1):75–98
26. Food Standards Australia New Zealand (2006) A risk profile of dairy products in Australia [cited 2016 Dec 12]. Available from: http://www.foodstandards.gov.au/code/proposals/documents/P296%20Dairy%20PPPS%20FAR%20Attach%202%20FINAL%20-%20mr.pdf
27. Food Standards Australia New Zealand (2005) Proposal P265 primary production and processing standard for seafood; final assessment report [cited 2016 Dec 12]. Available from: http://www.foodstandards.gov.au/code/proposals/documents/P265_Seafood_PPPS_FAR.pdf
28. New South Wales Food Authority (2012) Food safety risk assessment of NSW food safety schemes [cited 2016 Dec 12]. Available from: http://www.foodauthority.nsw.gov.au/_Documents/science/Food_Safety_Scheme_Risk_Assessment.pdf
29. New South Wales Food Authority (2012) Seafood safety scheme – periodic review of the risk assessment [cited 2016 Dec 12]. Available from: http://www.foodauthority.nsw.gov.au/_Documents/scienceandtechnical/Food_Safety_Scheme_Risk_Assessment.pdf
30. Food Standards Australia New Zealand (2014) Proposal P1014 – primary production and processing standard for meat and meat products [cited 2016 Dec 12]. Available from: http://www.foodstandards.gov.au/code/proposals/Pages/proposalp1014primary5331.aspx
31. Astridge K, McPherson M, Kirk M, Knope K, Gregory J, Kardamanidis K et al (2012) Foodborne disease outbreaks in Australia 2001–2009. Food Aust 63(12):7
32. Temporary food stalls: City of Sydney (2015) [cited 2016 Dec 12]. Available from: http://www.cityofsydney.nsw.gov.au/business/regulations/food-and-drink-businesses/temporary-food-stalls
33. Requirements for the operation of a Temporary Food Stall: City of Sydney (2011) [cited 2016 Dec 12]. Available from: http://www.cityofsydney.nsw.gov.au/__data/assets/pdf_file/0003/65892/Requirements-for-the-operation-of-a-Temporary-Food-Stall.pdf
34. New South Wales Food Authority (l2015) Food safety supervisors and training [cited 2016 Dec 12]. Available from: http://www.foodauthority.nsw.gov.au/retail/fss-food-safety-supervisors
35. Mornington Peninsual Shire Council (2015) Food safety for business [cited 2016 Dec 12]. Available from: http://www.mornpen.vic.gov.au/About-Us/Business-Economy/Business-programs/Food-Safety-for-Business
36. State Government of Victoria, Department of Health and Human Services (2013) About Streatrader [cited 2016 Dec 12]. Available from: https://streatrader.health.vic.gov.au/about/
37. State Government of Victoria, Department of Health and Human Services (2015) Food premises classification and registration [cited 2016 Dec 12]. Available from: http://www.health.vic.gov.au/foodsafety/bus/class.htm
38. Mornington Peninsula Shire Council (2013) Guidelines for temporary food premises [cited 2016 Dec 12]. Available from: http://www.mornpen.vic.gov.au/files/e8097825-d1d4-4cc1-9e56-a204009f2af0/Mornington_Peninsula_Shire_Guidelines_for_Temporary_Food_Premises.pdf
39. Australian Farmers Market Association (2015) Farmers market food safety guide [cited 2017 Jun 15]. Available from: http://farmersmarkets.org.au/wp-content/uploads/AFMA-Farmers-Market-Food-Safety-Guide-Aug-15.pdf

Chapter 9
Identifying Hazards and Food Safety Risks in Farmers Markets

Benjamin Chapman and Allison Sain

Abstract Outcomes of survey studies on food safety practices in farmers markets and outbreaks of foodborne illnesses associated with farm stands and farmers markets illustrate the need for a better understanding of actual food handling behaviors in these environments. Observational studies have identified actual practices that may increase a consumer's risk for foodborne illness and are consistent with risk factors for foodborne illness as identified by the Centers for Disease Control and Prevention. In order to mitigate risks in farmers market environments, vendors and managers must first recognize that regardless of the size of a business, foodborne illness can result if risky food handling practices are used. They must be able to identify risky practices and strategies for controlling hazards. Practice assessment tools developed as a result of survey studies for use as self-inspections of farms and markets and tools developed for use in gathering data in observational studies in farmers markets may be useful in helping vendors and market managers identify food safety hazards and develop and implement risk mitigation strategies to enhance safety of products sold in farmers markets.

Keywords Observational studies • Farmers' market practices • Hazard identification • Risk assessment • Practice assessment tools

B. Chapman (✉)
Department of Youth, Family and Community Sciences, North Carolina State University,
512 Brickhaven Drive 220E, Campus Box 7606, Raleigh, NC 27695-7606, USA
e-mail: benjamin_chapman@ncsu.edu

A. Sain
Department of Food, Bioprocessing and Nutritional Sciences, North Carolina State
University, Schaub Hall, 118, Raleigh, NC 27695-7606, USA
e-mail: sasmathe@ncsu.edu

© Springer International Publishing AG 2017   119
J.A. Harrison (ed.), *Food Safety for Farmers Markets: A Guide to Enhancing
Safety of Local Foods*, Food Microbiology and Food Safety,
DOI 10.1007/978-3-319-66689-1_9

Farmers' markets provide opportunities and benefits to local communities that other food retail venues do not. Farmers' markets improve economic viability of local communities by capturing and retaining a greater portion of farm and food dollars in local economies [1]. Farmers' markets are highly visible and intuitively obvious sites for food producers and patrons to find each other, while allowing producers to capture greater value for food products [1–3].

The U.S. Centers for Disease Control and Prevention (CDC) estimate that each year approximately one in six Americans gets sick, 128,000 are hospitalized and 3000 die of foodborne diseases [4]. The U.S. Food and Drug Administration (FDA) has identified five risk factors that increase the likelihood of a foodborne illness in retail settings [5]. These include:

1. improper cooking procedures;
2. temperature abuse during storage;
3. lack of hygiene and sanitation by food handlers;
4. cross-contamination between raw and ready-to-eat foods;
5. acquiring food from unsafe sources.

Studies with both self-reported practices and actual observations have determined that several of these risks are likely to exist in farmers markets [6–8].

Vendors at farmers markets sell all food commodities, but there seems to be a focused market on fruits and vegetables. A common misconception is that local fruits and vegetables grown on small farms are automatically safer than foods transported long distances from large "industrial" sized farms [9]. Considering the rising popularity of farmers markets and the increasing number of produce outbreaks, a focus on safety of food products sold at farmers markets can protect farmers, patrons, and local economies. In a study conducted by Park and Sanders, 1564 fresh samples of ten vegetable types were tested for thermotolerant *Campylobacter* [10]. Of the 1564 vegetable samples, 533 were from farmers markets and the remaining samples were from supermarkets. Six types of vegetables from the farmers markets were positive for *Campylobacter*, while all samples from supermarkets were negative. The study found that washing with chlorinated water removed the contamination and suggested that vegetables sold at farmers markets were produced and/or stored under less sanitary conditions than those sold at supermarkets. The authors concluded that these vegetables "could constitute health hazards" [10].

Poultry products are also sold in many markets. Scheinberg et al. compared the prevalence of *Campylobacter* and *Salmonella* in fresh and frozen chicken from farmers markets and supermarkets [11]. Chicken from farmers markets had significantly higher prevalence of both *Salmonella* and *Campylobacter* compared to conventional of organic chicken from supermarkets [11]. While microbial presence was recorded, a needs-assessment was conducted with farmers market vendors to analyze food safety knowledge and attitudes about poultry products. Twenty of 21 (95%) vendors who participated in the survey believed their poultry products sold through the farmers market sector were safer than conventionally produce poultry sold through commercial shopping centers [12].

In a study by Worsfold and colleagues, vendors surveyed (n = 50) self-reported their food hygiene standards as high, even though less than half had established risk management procedures, and most did not categorize their produce as high-risk [13]. While some farmers markets had access to electricity, handwashing stations, toilets, trash receptacles, and cleaning procedures, others were noted to have limited access to such facilities. The desires of farmers market patrons did not include food safety but instead focused on "freshness, traceability, taste, quality", and production method. Because of limitations such as inaccurate responses and lack of data due to self-reporting, this study provides a limited amount of information on the food safety culture at farmers markets.

A preliminary U.S. study used smartphone technology to record direct observations of food safety practices at an Indiana farmers market [14]. Observations of 18 employees engaging in 900 sequential food handling transactions "revealed that food safety behaviors were infrequently practiced, suggesting an increased risk of foodborne illness" [14]. Further analysis showed that the number of simultaneous work roles an employee was engaged in was positively associated with the number of potential violations, yet adding more employees to a booth did not ensure that work roles were properly segregated. This study demonstrates the importance of clearly defined work roles for employees for improving food safety at farmers markets [14].

Choi and Almanza compiled inspection scores from temporary foodservice establishments, including farmers markets, and restaurants to compare health inspection violations among these types of locations [15]. They compared 29 farmers markets and 120 full-service independent restaurants, among other types of establishments, and found that farmers markets had significantly fewer average numbers of food safety violations than full-service independent restaurants. However, they also noted that permanent establishments tend to have longer inspections and more categories to inspect and therefore a greater potential for violations [15]. Another limitation is that repeat violations were not considered; these could be a more important indicator of the enforcement of food safety practices. Additional factors that were not considered were the complexity of the menu and types of violations.

Self-completed surveys of 47 farmers market vendors at two farmers markets in Florida were collected as part of a study to better understand the need and areas for educational development [7]. Even though only 32% of the vendors had received food safety training, more than 50% of the vendors were very confident about their food safety practices. Simonne et al. concluded that the results from the surveys showed a need for the development of food safety training materials specific to the farmers market sector [7].

Unsafe Food Safety Practices on Farms and in Farmers' Markets Resulting in Foodborne Illness Outbreaks from Various Types of Products

2011 E. coli O157:H7 Oregon Outbreak (Strawberries)

In mid-July of 2011, the Oregon Health Authority started receiving reports of a foodborne illness outbreak [16, 17]. Fifteen individuals became ill between July 10 and July 29. Four individuals were hospitalized, two people suffered kidney failure, and one elderly woman died due to kidney failure, a complication associated with severe symptoms of E. coli O157:H7. Common food products and places visited were used to construct a trace back to the source.

Early in the investigation, officials encouraged patrons to not eat strawberries from any farmers markets in an attempt to reduce further illnesses. This caused major profit losses for vendors selling strawberries and other vendors in the farmers markets of Oregon. Through interviews with those who had become sick, the Health Authority was able to establish a connection between the E. coli O157:H7 strain and strawberries harvested from a local strawberry farm that were sold through multiple farm stands. Health officials began to gather a list of vendors who sold these strawberries at farm stands and farmers markets.

The farm had just finished their strawberry season when they were approached by the Health Authority for inspection and investigation to determine if the connection established through interviews could be confirmed through laboratory tests. Since the strawberry season was over, the only samples left to analyze were those preserved either through canning or freezing. These samples were received from patrons who had bought the strawberries at farm stands or farmers markets. The farm stands and farmers markets would buy strawberries directly from this farm and resell them to patrons.

Once the strawberries from the suspect farm were confirmed as the source of the infectious strain by laboratory tests, the Oregon Health Authority searched further on the farm to identify the source of contamination. Laboratory samples were taken from around the farm, including deer droppings. The deer droppings were confirmed as the source of the E. coli O157 that was present on the contaminated strawberries [16, 17]. This outbreak gives a prime example of contamination issues that can occur in small-scale farming and many farmers markets. If the grower is not ensuring risk reduction actions on the farm, such as fencing or other forms of keeping fecal contamination out of the fields, the grower is not conducting best food safety practices and someone can get sick.

2010 Iowa Salmonella Outbreak in (Guacamole)

In the summer of 2010, 44 cases of Salmonella Newport were linked to products sold at two Iowa farmers markets [18, 19]. Through epidemiological interview questionnaires and testing of individuals and implicated food products, it was determined that guacamole-based products produced by a local supermarket/ taqueria restaurant were implicated as the source. The restaurant sold guacamole; red and green salsa; and pork, chicken, and vegetable tamales at the farmers markets.

According to Iowa health officials, Linn County Public Health had inspected the restaurant stand at the farmers market on the implicated date. They found that some of the ice used for cooling had melted, increasing the potential risk for temperature abuse of the products. Officials identified factors associated with contamination such as lack of sanitation techniques, cross-contamination, and improper washing of avocados at the restaurant during preparation. Improper holding temperatures at the market were listed as a potential factor for bacterial increase because the temperature for that day was above 80 °F (26.7 °C) [18, 19].

2002 E. coli O157:H7 Outbreak (Unpasteurized Gouda Cheese)

Thirteen individuals became sick in 2002 with *E. coli* O157:H7 in Edmonton, Alberta, Canada, including two children who developed hemolytic uremic syndrome [20]. After environmental health officers conducted interviews and gathered clinical information on the sick individuals, pulsed field gel electrophoresis confirmed unpasteurized Gouda cheese from a local dairy as the source of *E. coli* O157:H7. Of the 13 individuals who became ill, eight had consumed the cheese at home, five had purchased the cheese at a local farmers market, two received the cheese as a gift, and one purchased the cheese directly from the dairy [20]. This outbreak is important for understanding that, while food products may be sold by a small number of vendors, multiple people can consume the product therefore exposing a relatively large number of people to unsafe product.

1994 Ontario Salmonella Outbreak (Soft Cheese)

In September of 1994, a woman notified the Waterloo, Ontario Regional Community Health Department of implicated cheese after she and her aunt consumed a soft spreadable breakfast cheese at a local farmers market [21]. The two individuals suffered symptoms of foodborne illness including diarrhea and thought that the source may have been the cheese product. One of the women received medical attention and through the culture of her stool sample, *Salmonella* Berta was identified. The implicated cheese product tested positive for *Salmonella* Berta and *E. coli*. During further investigation, it was found that the cheese was produced at a licensed, local dairy using unpasteurized milk. The process of making the cheese was investigated to show further insight into the production of the cheese product.

Time and temperature seemed to be a potential source of risk with the production of the cheese product. The skim milk curds were ripened at room temperature for 23 days, and the temperature and time were not monitored extensively. Conversations unveiled that the farmers used the same buckets used for ripening the cheese to soak raw chicken carcasses. The farmers recounted that the buckets were most likely not disinfected properly prior to the cheese production. The farmers disclosed that two of them, and later another family member, had become sick. The potential for cross-contamination and improper handwashing were also identified as potential risk factors. Through investigations, 82 cases of *Salmonella* Berta (35 confirmed, 44 suspected, three secondary) were associated with this particular outbreak [21].

Observation of Food Safety at Farmers' Markets

Previous food safety research has been conducted in the farmers market sector [7, 13], but observational research of the farmers market food safety culture data in order to develop an educational curriculum is lacking. In order to better understand the current food safety practices and knowledge of farmers market vendors, food handling and risk reduction practices were observed and farmers market managers were surveyed in a study of North Carolina (NC) farmers markets [8].

In the spring of 2010, 20 secret shoppers were trained to collect quantitative and qualitative observational data. The secret shoppers were North Carolina County Extension Agents selected based on their close relationship with the farmers markets in their county. The training covered: (1) food safety basics such as factors contributing to foodborne illness and outbreaks in the farmers market sector, (2) background on food safety needs of farmers markets, (3) goals of the project, and (4) defined specific questions on the observational instrument.

Secret shoppers were trained to collect quantitative and qualitative observational food safety behavior and facility data on the farmers markets through the use of an observational instrument (Fig. 9.1). During the development stages of the observational instrument, risk factors contributing to foodborne illness were used to identify potential risk factors in the farmers market sector. FDA risk factors, coupled with five systematic visits to operational farmers markets, were used to develop the observational instrument. This observational tool was designed for use by secret shoppers to collect data incognito to ensure the data was a representative sample of actual behaviors and facilities at farmers markets. After receiving training on the observational tool, secret shoppers were assigned farmers markets outside of their counties to attend for a one-day visit. Data collectors visited 37 markets; these markets were selected through convenience [8].

Farmers' Market Self Assessment

Location: _____

Date & Time: _____

QUESTIONS	YES	NO	COMMENTS
Overall is the market clean?			
Are the vendors under a permanent awning or protective covering, or do individual vendors have temporary structures (tents) set up at the market?			
Are different products separated into designated areas? eg: produce area/seafood area/non-food area			
Is anyone smoking around the food products?			
Is there a concession stand at the market? If there is a grade posted, what is it?			
Are there restrooms in the market?			
Are the restrooms clean?			
Are the restrooms in goodworking condition?			
Are there displays in the restrooms of proper handwashing techniques?			
Are there handwashing facilities inside/ outside the restroom operatioonal?			
Is soap provided?			
Is there a posted cleaning schedule, if so when was it last updated?			
Is there bird netting under the awning?			
Are there signs of rodents? eg: birds/rats			
Are live animals (eg: dogs) allowed in the market area? Do vendors or patrons have animals with them?			

Fig. 9.1 Farmers' market assessment form used in observational study of North Carolina farmers markets. Courtesy of Ben Chapman

Foods Available

1. Top 3 Foods: What are the 3 main foods that the vendor sells? (Mark 1 through 3, with 1 being the main food)
2. Percentage: For the Top 3 Foods, what percentage of the vendor's total product do each compose?
3. Check: For all other products the vendor sells, place a check mark beside the product name.

Baked Goods		Cucumbers	
Canned Foods		Grapes	
Condiments/Spices		Greens	
Dairy		Melons	
Eggs		Nuts	
Fish/Seafood		Okra	
Frozen Foods		Peaches	
Herbs/Spices		Peppers	
Juices		Potatoes	
Meats		Sprouts	
Apples		Squash	
Beans		Strawberries	
Blueberries		Tomatoes	
Broccoli		Watermelon	
Cabbage		Refrigerated Foods	
Corn		Other.	

Questions to Ask Vendors

A. Are there any organic or synthetic chemicals/fertilizers/pesticides/manures used on the products?

B. Are all foods grown/processed by the vendors? If not, how do they assess safety of their supplier

C. If yous ee signs with claims such as "clean" or "washed" what does this mean?

D. How are the foods transported to the market? eg: refrigerated/closed storage

E. What kind of soil were the products grown in? eg: organic/compost/plant material

F. What risks do the vendors worry about? Do they mention GAPs? If not, prompt them

Answer:

Fig. 9.1 (Continued)

▮▮▮▮▮▮▮▮▮▮▮▮▮	Do the employees appear clean? (eg: do they have dirty hands? Are their clothes dirty?)
How many people are working? (circle one)	
1 2 3 4 5+	
How far are products from the ground? (circle one)	
directly on the ground not on the ground	
Are tables covered? What material? eg: cloth/grass carpet/wood/plastic	Are single-use gloves used?
What material are the produce containers? eg: wood/cardboard/plastic	Is there a designated person in charge of money transactions?
Is there a thermometer visible?	Are products transported in plastic lugs/totes or cardboard boxes, are those boxes reused? Are they labeled with the vendors farm name?
Are foods in direct contact with ice?	Are there any signs displaying good food safety practices?
Are there samples? What are they? Are they covered/on ice?	

Other Comments:

Fig. 9.1 (Continued)

Restrooms, Handwashing Facilities, and Vendor Health and Hygiene

Fifty-four percent of the farmers markets in the NC study provided restroom facilities in or near the market [8]. Based on the six categories of: (1) poor air quality (odors), (2) alterations/obstructions (broken), (3) filth (trash), (4) possible contamination (dirt), (5) toilet seats with matter on them, (6) spills of urine or fecal matter around the toilet and sink, 73% of the available restrooms were categorized as clean. Forty-nine percent of the farmers markets provided handwashing facilities in the market. However, 27% of farmers markets did not have any access to handwashing facilities in or around the market. In the available handwashing facilities, 87% did not display graphics showing proper handwashing techniques, 22% of the handwashing facilities did not have soap available, and only 11% of restrooms had a cleaning schedule posted.

The visible appearance and actions of vendors were observed to provide a better understanding of their health and hygiene. Some vendors were seen in clean clothes, but others were seen in clothes that obviously were worn on the farm previously and not washed prior to attending the farmers markets. Vendors were seen eating without proper handwashing before helping patrons to package food products for purchase. Vendors were seen using single-use gloves for produce pickup, and vendors also were observed using single-use gloves for multiple tasks such as food sample preparation, money handling, and wiping the face or personal clothing. One vendor provided handouts on cooking suggestions and safe food handling tips from the USDA. Hand sanitizer was seen in some vendor stands for use only by the vendor, while in other stands, hand sanitizers were located where patrons could also use the sanitizer [8].

Types of Products Available for Purchase

At the farmers markets observed, 82% of the food products for purchase were produce. Potentially hazardous food products such as melons, leafy greens, and sprouts were seen for purchase [8].

Production Attributes, Practices, and Transportation of Food Products

Since production practices were not observed as part of this study, they were better understood through conversations with vendors as well as through observations of product labeling. They were assessed through indicators such as information contained on product labels and conversations with the vendors about the location of their farm or processing unit, soil, water, irrigation, compost, pesticides, and fertilizer used with products, if the vendor was selling produce. Production attributes of meat producers selling at the famer's markets included: free range, pasture fed, rotational grazing in pastures, corn feed, grass feed, no antibiotics, and growth hormone free. Produce farmers selling at the farmers markets used a variety of descriptors to describe their production practices. These included non-organic and organic, certified and non-certified organic, USDA Organic, no additives or preservatives, certified chemical free and fertilizer free. Other practices described included the conventional use of chemicals and pesticides, sprayed when necessary, use of unpasteurized manure, use of worm compost, use of herbicides and use of fish emulsion. Based on conversations the secret shoppers had with the vendors, products sold were grown or produced by the vendor, by a friend or neighbor of the vendor, and sometimes by an unknown third party. Most of the food products were grown or produced in North Carolina, but products from Georgia were observed for sale. Foods were observed transported in different vehicles and/or equipment of varying visual-cleanliness, including open and enclosed trucks, cold storage, horse trailers, electric iceboxes, and trunks of cars. During conversations with vendors selling produce, 85% reported they were unaware of Good Agricultural Practices (GAPs), or were at least unresponsive when prompted. The few vendors who understood the implications of GAPs believed the methods were irrelevant to the farmers market sector [8].

Food Sales and Sampling

Food samples were seen of products ranging from soft cheeses, fried cheeses, pepper jelly, herb-based dips, blueberries, edamame, fresh cut apples, fresh cut melons, and cakes. These samples were witnessed covered and uncovered; on ice and not on ice; cut with non-food grade pocketknives and food-grade kitchen knives; distributed with forks, toothpicks, and disposable cups, handled with single-use gloves and handled with bare hands. Potentially hazardous foods such as fresh cheeses and cut melons were observed stored without ice and refrigeration. Only one thermometer was observed being used during food product storage. Products such as corn, watermelon, cucumbers, potatoes, and squash were seen placed on the ground for display and purchase. Products were also seen stored on grass carpet, in plastic lugs, on burlap, in wooden baskets, in paper bags, and in trash bags [8].

Manager Surveys

Following secret shopper activities, a survey was conducted with the manager of each of the 37 visited farmers markets [8]. Managers were contacted through phone surveys and informed that all information provided would be used to develop educational materials specifically created for the farmers markets of North Carolina. The survey questions helped confirm observations collected by the secret shoppers. Managers were asked questions about their market that focused on: (1) vendor guidelines and products allowed for purchase, (2) facilities available and hygiene process used for cleaning, (3) major challenges, (4) food safety materials, training, and/or policies provided or needed for vendors, and (5) the best method of delivering educational materials. Managers were made aware that the purpose of the surveys was to better understand the functions of a farmers market, guidelines set for the vendors, and the importance of food safety information for the market.

Information collected from surveys with managers of 28 of the 37 (76%) farmers markets observed provided a greater understanding of market operations and guidelines. Markets involved in this study include markets operated by different organizations and had been in operation for 3–32 years. Ninety-three percent of the markets permit the sale of all food commodities, but 7% of markets limit sales to agricultural products. Most markets accept vendors who produce food conventionally, and some have vendors who are GAPs certified and/or organic certified. Some markets require farm/kitchen inspection by the manager or market council prior to the vendor being accepted into the market, whereas other markets do not have any guidelines for who sells at the market. Some markets provided a radius in which food products must be grown or produced; other markets did not have location requirements. Markets were recorded to occur in parks, permanent open-markets, under tents, in city halls, and parking lots allowing for some markets to have limited to no access to running water, restroom facilities or electricity. Market meetings were reported to occur weekly, monthly or yearly, according to market desire [8].

When prompted to list the major challenges of a farmers market, only one manager referenced food safety; managers listed weather, market location, increasing number of patrons, and publicity as the main challenges of the farmers market. Six market managers felt that food safety was not as great of a concern in the farmers market sector as compared to other food retail venues such as grocery stores; comments usually included references to increased freshness and cleanliness due to decreased travel, increased washing, and decreased pesticide residues.

During conversations about food safety resources and methods, managers requested specific guidelines for activities that vendors should do at the markets in regards to food safety. Sixty-eight percent of managers (25 out of 37) believed the most effective method to communicate food safety to be brochures and Internet resources. Managers requested that Internet resources not be the only venue because some of the vendors in the markets do not have access to computers in their homes. Some managers desired to provide meetings on food safety led by either the manager or an Extension Agent. All of the managers surveyed felt that their market was not exempt from food safety education [8].

Next Steps

The recorded risk factors define the need to promote education on food safety behaviors and best risk reduction practices and provide a foundation on specific areas in which vendors should be trained. Considering GAPs are the most relevant safe practice guidelines for vendors who sell produce direct to consumers at farmers markets, relevant guidelines should be developed and communicated to members of farmers markets in order to increase the safe food practices by vendors [7]. New guidelines created specifically for risk management at farmers markets will aid in the reduction of food safety risk actions.

Low availability of handwashing facilities and poor food handler hygiene indicate a potential for contamination of food and food surfaces in these markets. The facilities available to vendors were similar to those reported in other market studies [13]. FDA has identified lack of hygiene and sanitation by food handlers as one of the risk factors that increases the likelihood of a foodborne illness [5]. Even though practices of handwashing were not recorded in the North Carolina study, based on the low availability and maintenance of handwashing facilities, a high risk for poor hygiene and low adherence to handwashing by food handlers was conveyed. Research supports that the unavailability or inaccessibility of handwashing stations decreases the adherence to hand hygiene [22, 23]. Education on the need to have available and easily accessible handwashing stations is needed. Also, education or training activities that increase food handler knowledge, perception, and attitude towards handwashing is needed [24, 25].

Improper time-temperature controls during food preparation, storage and transportation for foods to be served as samples at the point of purchase were observed. These findings are similar to other research that looked at food handler knowledge and behaviors during food preparation [26, 27]. The FDA has identified temperature abuse during storage and improper cooking procedures as risk factors that increase the likelihood of a foodborne illness [5]. Products needing to be held at lower temperatures, such as cut cantaloupe, were observed stored without temperature controls such as refrigeration or ice packs. Also, considering only one thermometer was observed for food storage, this shows a lack of monitoring time-temperature control for safety food products for sampling and for purchase. Other research has identified barriers to why food handlers do not use or rarely use thermometers, including because food handlers believe thermometers to be a time constraint, they lack a working thermometer, they do not know the temperature to look for, and they do not know how to take temperatures [28]. These barriers as well as the importance of keeping foods out of the danger zone (temperature range between 41 °F and 135 °F or 5 °C and 57.2 °C) need to be communicated in training materials.

The manager surveys provided confirmation that North Carolina farmers markets need a food safety curriculum that is tailored to manager and vendor practices and facility need. The necessity of food safety education materials developed specifically for the farmers market sector is supported by Simonne et al. whose study looked at the need and areas for educational development in two Florida farmers markets [7]. Considering the high variability in farmers market facilities, locations, guidelines, vendor types, and resource need, the curriculum will need to be specific to food safety risks but allow for vendors and managers to assess the needs of their individual markets to provide specific risk reductions to their markets. Tools presented in Figs. 9.1 and 9.2 can help vendors and managers identify food safety risks in a market.

Good Farmers' Market Practices (GFMPs)

As part of a good food safety culture, market managers and vendors need to know the risks associated with the products or meals produced, know why managing the risks is important, and how to effectively manage potential risks.

Food safety focuses on the handling, preparation, and storage of food in ways that reduce the risk of foodborne illness. Since microorganisms cannot be seen without a microscope, preventing contamination, destroying harmful microorganisms, and limiting the growth of harmful bacteria are steps to reduce the risk of foodborne illness.

GFMPs must include the following areas:

1. Food safety principles
2. Health and hygiene practices
3. Providing safe samples.

Food Safety Principles for Market Managers and Vendors

Important factors to note that contribute to the risk of foodborne illness, in no specific order, and why these should be controlled include the following [5].

1. Improper hot and cold holding temperatures of time-temperature control for safety (TCS) foods. The purpose of holding TCS foods at proper temperatures is to minimize the growth of any pathogenic bacteria that may be present in the food. TCS foods that are going to be held at cold temperatures must be held at a temperature of 41 °F (5 °C) or below. TCS foods that are going to be held at hot temperatures must be held at a temperature of 135 °F (57.2 °C) or above. Food facility operators must take every precaution to minimize the amount of time that time-temperature control for safety foods spend in the danger zone (temperatures between 41 °F and 135 °F or 5 °C and 57.2 °C).
2. Improper cooking temperatures of foods. Cooking food to the proper temperatures is extremely important because many raw meats and other non-ready to eat foods have pathogenic bacteria on them naturally.

Farmers Market Self-Help Form

$\mathcal{P}$ractice	YES	NO	DOES NOT APPLY TO ME
Training, Certifications & Requirements			
Our market has established food safety standards or rules.			
Our market requires documentation of certification or evidence of food safety training from producers in order to sell in the market.			
Our market has evidence of inspection by the state Department of Agriculture.			
Vendors selling eggs have a Department of Agriculture license or candling certificate.			
Meat sold at the market is ONLY from U.S. Department of Agriculture certified slaughter facilities and follows compliance laws and/or follows state guidelines for my state.			
Meat is sold only from a refrigerated unit that meets state Department of Agriculture guidelines.			
We require vendors who sell milk to sell only milk that has been pasteurized.			
Land & Water Use			
We ask farmers/vendors how their land was used before production of crops they are selling and whether efforts have been made to prevent contamination.			
We ask farmers/vendors about their use of manure on food crops.			
We ask farmers/vendors whether they limit domestic and wild animals in food production areas.			
We ask farmers/vendors if crop production areas are near or adjacent to animal production areas or possible run-off from these areas.			
We ask farmers/vendors about the sources of irrigation water used on the crops.			
We ask farmers/vendors if the water used in crop production is tested for bacteria or if it is from a municipal supply.			
We ask farmers/vendors if the water used to rinse produce on the farm is tested for bacteria or if it is from a municipal supply.			

Fig. 9.2 Enhancing the Safety of Locally Grown Produce—Farmers' Market Self-Help Form. Courtesy of Judy Harrison. University of Georgia Publication #FDNS-E-168-11

$\mathcal{P}$ractice	YES	NO	DOES NOT APPLY TO ME
Workers at the Farm/Production Site			
We ask farmers/vendors if they have policies that limit sick workers from coming in contact with the products they are selling.			
We ask farmers/vendors if they provide sanitation training for their workers.			
We ask farmers/vendors if they provide training for their workers on proper glove use.			
We ask farmers/vendors if they train workers to seek immediate first aid for injuries like cuts, abrasions, etc. that could be a source of contamination for produce or other food products.			
We ask farmers/vendors if they train workers on what to do with products that come in contact with blood or other bodily fluids.			
Facilities & Equipment at the Farm/Production Site			
We ask farmers/vendors if their workers have easy access to handwashing facilities with clean water, soap and paper towels in or near the fields or production areas.			
We ask farmers/vendors if their workers have easy access to handwashing facilities with clean water, soap and paper towels in or near the packing area or processing area.			
We ask farmers/vendors if workers have easy access to toilet facilities in or near their fields, but at a safe distance to prevent contamination.			
We ask farmers/vendors if workers have easy access to toilet facilities in or near their packing areas or processing areas, but in an area that minimizes risk of contamination.			
We ask farmers/vendors if harvesting equipment (knives, pruners, machetes, etc.) is kept reasonably clean and is sanitized on a regular basis.			
We ask farmers/vendors if harvesting containers and hauling equipment are cleaned and sanitized between uses.			
We ask farmers/vendors if surfaces that come in contact with fruits, vegetables or other foods at the production or processing site are cleaned and sanitized often and regularly.			
We ask farmers/vendors if damaged containers are properly repaired or discarded.			
We ask farmers/vendors if cardboard boxes used are new and used only once.			
We ask farmers/vendors if precautions are taken to keep glass, metal, plastic fragments, rocks or other dangerous or poisonous items away from produce or other food products.			
Storage & Transport by the Farmer/Vendor			
We ask farmers/vendors if containers used with fruits, vegetables and other foods are cleaned between each use.			
We ask farmers/vendors if fruits and vegetables are cooled after harvest.			

Fig. 9.2 (Continued)

$\mathcal{P}ractice$	YES	NO	DOES NOT APPLY TO ME
We check to see that fruits and vegetables arrive cool at market.			
We ask farmers/vendors if the vehicle used to transport fruits and vegetables or other foods to market is cleaned between each use.			
We look to see if trucks or other vehicles bringing in food to the market appear to be clean.			
We ask if farmer/vendors could trace the food back to exactly where it was produced on the farm and in the packing/processing facilities.			
Health and Hygiene of Vendors/Market Workers			
We train the vendors in good sanitation practices.			
We train our market workers on proper handwashing.			
Workers in our market appear to be clean.			
We have a policy that limits sick workers from coming in contact with foods sold in the market.			
We train vendors on safe handling of foods offered as samples at the market.			
We train market workers on safe handling of foods that can be sampled at the market.			
We train vendors on sanitary procedures for allowing sampling at the market.			
We train market workers on sanitary procedures for allowing sampling at the market.			
Market Facilities and Equipment			
Toilet facilities are easily accessible at the market.			
Toilet facilities are serviced and cleaned on a regular schedule.			
Handwashing facilities are easily accessible by workers at the market.			
Handwashing facilities at our market are cleaned and stocked with clean water, soap and paper towels on a regular schedule.			
If no handwashing facilities are available at the market, we do at least provide hand sanitizer in our market.			
Conditions in the market stalls appear to be clean.			
Cardboard boxes used to hold produce are removed and discarded as they become empty.			
Cardboard boxes are NOT reused to hold new supplies of produce to be sold.			
Produce is displayed in plastic containers that appear to be clean.			
We do NOT allow wooden crates to be used for produce.			

Fig. 9.2   (Continued)

Practice	YES	NO	DOES NOT APPLY TO ME
Coolers or refrigerator units at the market are cleaned and sanitized on a regular basis.			
We take precautions to prevent products in refrigerators, coolers, etc. from becoming contaminated (covering, packaging, changing air filters, etc.).			
We take precautions to keep raw meat products separate from ready-to-eat foods like fresh produce, etc. at the market.			
We clean and sanitize surfaces on a regular schedule at the market that come in contact with food products being sold.			
We use food thermometers to check temperatures of foods in coolers/refrigerators at the market.			
We use food thermometers to check temperatures of foods prepared for sampling in the market.			
We calibrate the markets' food thermometers regularly to make sure they are reading accurately.			
There is evidence of vendors using thermometers to monitor holding conditions of products being sold.			
We keep cut fruits and vegetables at refrigerator cold temperatures (41°F or colder) for sampling (either in refrigerator, cooler or on ice).			
Perishable produce like lettuce is displayed on ice, in coolers with ice or are kept refrigerated.			
We keep hot foods for sampling at 135°F or higher.			
Customers			
Handwashing facilities are easily accessible by customers at the market.			
We do not have handwashing facilities at the market, but we do provide hand sanitizer for customers.			
Our market enforces a policy of "no pets" allowed in the market.			

If you answered "no" to any of the questions, those questions represent areas where changes or improvements may help your market to offer safer products and reduce potential risk of foodborne illness. Please read the *Enhancing the Safety of Locally Grown Produce* factsheets for your risk area to learn how to minimize risk.

This project was supported all, or in part, by a grant from the National Institute of Food and Agriculture, United States Department of Agriculture (Award Number 2009-51110-20161).

Publication #FDNS-E-168-11. J.A. Harrison, J.W. Gaskin, M.A. Harrison, J. Cannon, R. Boyer and G. Zehnder.   February 2012

The University of Georgia and Ft. Valley State University, the U.S. Department of Agriculture and counties of the state cooperating. Cooperative Extension, the University of Georgia Colleges of Agricultural and Environmental Sciences and Family and Consumer Sciences, offers educational programs, assistance and materials to all people without regard to race, color, national origin, age, gender or disability. An Equal Opportunity Employer/Affirmative Action Organization, Committed to a Diverse Work Force.

Fig. 9.2 (Continued)

3. Contaminated utensils and equipment. When utensils or equipment become contaminated, they can transfer that contamination to the food that could result in a foodborne illness.
4. Poor employee health and hygiene. It is imperative that food workers are in good health while preparing food. A food worker that has been diagnosed with an acute gastrointestinal illness (GI), or is showing symptoms such as diarrhea, or vomiting in conjunction with diarrhea, could potentially contaminate food. It is possible for a food worker to transfer their illness to customers via the food. Even more disconcerting, there is the potential for employees working with large batches of food to spread the illness to numerous people causing an outbreak with high numbers of cases.
5. Food from unsafe sources. Any food that is to be sold, served, given away, or used as an ingredient, must be obtained from an approved source. An approved source is a facility where the food produced, prepared, or processed, meets or exceeds established food safety standards for that product.

Certain high-risk populations are not only at risk of contracting a foodborne illness but may also suffer more severe symptoms.

• Young children are at a greater risk of foodborne diseases because their immune systems are still developing and the protection afforded by the resident gut microflora is not as effective as in adults. Children are also more vulnerable to the toxic effects of chemical contaminants in foods.
• The elderly are more susceptible to foodborne illness than other groups because defenses or the ability to fight diseases decrease with age. A decrease in stomach acid secretion, which is a first line defense against ingested bacteria, compounds the problem.
• Pregnant women are naturally more susceptible because they are immunocompromised as a consequence of pregnancy.
• Due to the weakened immune systems, individuals suffering from cancers and chronic illness such as HIV/AIDS are particularly prone to contracting foodborne illness.

It is important for managers and vendors to understand that the factors that allow for bacterial growth are important, and all play into the safety of a food product:

• Temperature: Bacteria do not grow well when the temperature of the food is colder than 41 °F (5 °C) or hotter than 135 °F (57.2 °C). Keep time-temperature control for safety foods out of the Danger Zone! When food is left in the Danger Zone, bacteria can multiply quickly, and some create toxins.
• Oxygen: One important cause of food spoilage is air and oxygen. Oxygen can provide conditions that enhance the growth of many microorganisms. Some bacteria require oxygen for growth (aerobes) while others can grow only in the absence of oxygen (anaerobes). Many bacteria can grow under either condition and are called facultative anaerobes as discussed in more detail in Chap. 1. Most yeasts and molds that cause food to spoil require oxygen to grow. They can often be found growing on the surface of foods when air is present. Canned and vacuum

packaged foods provide an anaerobic environment. While canning provides a great method for preservation, knowing the correct method and recipe for canning is a must. Improper canning can lead to sickness or even worse, death from the potential growth of the anaerobic bacterium, *Clostridium botulinum.*

- pH: The acidity of foods has been used for centuries to preserve foods. Acidity plays a primary role in the preservation of fermented foods and combined with other factors such as heat, water activity, and chemical preservatives acts to prevent food deterioration and spoilage. The intensity of acidity of a food is expressed by its pH value as presented in Chap. 1. The pH of a food is one of several important factors that determine the survival and growth of microorganisms during processing, storage and distribution. Consequently, food processors are interested in determining the pH of foods and in maintaining pH at certain levels to control microbial growth and prevent product deterioration and spoilage. The pH scale runs from 0 to 14. The values that are less than 7 are acidic, while those greater than 7 are alkaline. A pH value of 7 is neither acid nor alkaline and is considered neutral. Most pathogens are controlled below a pH of 4.6.
- Water activity: Water in food that is not bound to food molecules can support the growth of bacteria, yeast, and mold. The term water activity refers to this unbound water. Pathogens cannot grow, or produce toxin in foods with a water activity below 0.85. The amount of available moisture can be reduced usually by the addition of high concentration of sugar or salt which bind free water or by dehydrating the food.

Health and Hygiene

It is a best practice to have restrooms with handwashing facilities available in a location where people will use them. If the facilities are too far away from that market location then most people will not visit the facilities. A handwashing station should be available in the restroom because hand sanitizer is not effective against viruses so it should not be used in place of handwashing. Also, considerations should be made to ensure that everyone has access to the handwashing stations that are available. Restrooms are not to be used for storage of food, equipment, or supplies. Toilet facilities should be clean and separated from other areas. Toilet paper should be provided in a permanently installed dispenser at each toilet. Handwashing facilities should be separated from toilets.

The FDA recommends handwashing facilities to be at a convenient distance of food production [29]. Vendors who provide samples are encouraged to have a handwashing station in their booth. Sinks should be of appropriate height and clearly marked. Handwashing signs should be in English as well as in other languages for patrons who may not speak primarily English.

Handwashing stations should include:

- Water source
- Soap

- Clean water
- Paper towels
- Catch basin for wastewater
- Trash receptacle.

While gloves can be helpful for avoiding bare-hand contact with food, they must be used for single task and discarded if damaged or soiled. Market managers and vendors should ensure food handlers are not ill and promote effective handwashing by food handlers. Avoid bare hand contact with ready-to-eat foods with the use of gloves, tongs, spoons, and hand papers (deli tissues). Clothing can be a factor in pathogen transfer so encourage those in attendance at the market to wear clean clothing, not the same clothing that they have worn while working on their farm. Workers can carry microbial pathogens on their skin, in their hair, on their hands, and in their digestive systems. Unless workers understand and follow basic food protection principles, they may unintentionally contaminate fresh and fresh-cut produce, food contact surfaces, water supplies, or other workers, and thereby, create the opportunity to transmit foodborne illness.

Providing Safe Samples

Best practices for providing samples include using equipment, cutting boards, knives, and utensils that are easy to clean, in good condition, and free of cracks. Protect equipment from contamination during transit to market and when not in use. Cutting boards should only be placed on sanitized food-contact surfaces. Cutting boards can be a reservoir for contamination. Do not use the same cutting board for raw foods and ready-to-eat foods. Food contact surfaces should be cleaned with detergent and water and then sanitized with either chlorine or quaternary ammonia-based sanitizers.

Air-drying food contact surfaces is best or wiping with a single-use paper towel. Sanitizers exposed to air lose concentration over time and should be made fresh periodically when used as dips. Sanitizers placed in spray bottles hold concentration for longer periods but should be made fresh at least daily. Cleaning and sanitizing chemicals should be stored away from food products. Food items should be kept at least six inches (15.24 cm) above the ground. Samples should be covered to protect from insects, dust, and other contaminants. Prevent patrons from touching samples.

Cook all time-temperature control for safety foods at approved temperatures for the required duration, and keep them out of the temperature danger zone while holding them for serving.

When time alone is used as a safety control for samples being served at a farmers market, it is safest to use the "Two Hour Rule" for consumers recommended by USDA [30]. This rule states to discard TCS foods if they have been out of temperature control for two hours or one hour, if the surrounding temperature is above 90 °F (32.2 °C). Outdoor farmers markets in certain areas may reach temperatures in excess of this temperature during hours of operation. An accurate thermometer should be used to measure sample holding temperatures. "Fresh-cut produce" is fresh fruits and vegetables for human consumption that have been minimally processed and altered in form by one of the following methods: peeling, slicing, chopping, shredding, coring, or trimming. These processes can be carried out with or without washing, prior to being packaged for use by the consumer. To help reduce the risk of foodborne micro-organisms, fresh-cut produce should be prepared fresh every morning of the farmers market. Using a first-in, first-out approach so as to not mix the new produce with the old produce is also a best practice.

Cross-contamination is the physical movement or transfer of harmful bacteria from one person, object or place to another. Preventing cross-contamination is a key factor in preventing foodborne illness. Good food safety practices that aim to reduce the risk of cross-contamination include keeping raw food and prepared food separate. Clean and sanitize all surfaces such as cutting boards and counters between raw and ready-to-eat food preparation. Use different utensils such as knives, tongs and lifters for raw and ready-to-eat foods, if cleaning and sanitizing between uses is not practical. Hands contaminated with meat juices can be great vehicles for cross-contamination. While storing food prior to travelling to market or at the farmers market, try to limit the amount of handling a food product receives. This will help reduce the risk of cross-contamination.

Using food-grade containers to store food products is also necessary to reduce illness risks. While garbage bags may seem like a good method for transporting a large quantity of food, some are treated with mold inhibitors and other chemicals. Therefore garbage bags are not food-grade.

Ice has important food safety considerations and should only be made with potable water. Use of clean, non-breakable utensils is a best practice to handle ice, such as tongs or an ice scoop. Avoid touching ice with dirty hands or glasses. Ice should be stored only in clean containers that are safe for storing food. It may be more economical to use cold packs as opposed to ice, and cold packs will prevent food from becoming overly wet or soft. The ice that is used to keep food products cold should not be served or sold for human consumption.

Waste containers should be available throughout the market for everyone to use and should be covered and emptied often. It is recommended that each vendor supplying samples provide a small garbage can for patrons to discard their used sampling containers or utensils.

Conclusions

While numerous studies have been published that examine many aspects of farmers markets, they do not include any that look at management practices, food safety plans, responses to foodborne illness outbreaks and recalls, microbiological analysis of potentially hazardous foods, or even the connection among any of the afore-mentioned items. Looking forward, researchers could compare on-farm procedures with those in the markets, perform risk analyses of farms and kitchens of vendors, examine whether formal food safety plans are effective in lessening risk of foodborne illness, look at the transportation methods used with farmers market items, or even the infrastructure of markets themselves.

Small or large businesses are not inherently good or bad when it comes to microbial food safety, nor are those who sell into a distribution chain any more or less safe than those who direct market their wares. Food safety hazards can exist in either environment. A small producer, growing tomatoes, leafy greens and herbs can reduce risks just as effectively as a large producer with millions of dollars in annual sales.

What matters more than size or market is whether the producer recognizes hazards and puts steps in place to reduce the risk of foodborne illness. Every business and organization fits somewhere on a continuum between positive and negative food safety culture. The quick tests for where they lie are: can everyone in the business recognize risks they are responsible for limiting, have all the tools they need to do so, and actually do it. An outbreak can affect the entire farmers market brand.

Summary

Risk factors identified by FDA as resulting in risk of foodborne illnesses have been observed in farmers markets. It is necessary to learn from past outbreaks and address factors that increase risk. Implementing preventive guidelines, for all food producers, handlers, and vendors, is necessary not only to protect patrons, but also to protect vendors from potential scrutiny and liability, inability to sell products, and loss of profit. These include providing well maintained and well-stocked bathroom and handwashing facilities for vendors and customers; practicing good personal hygiene and cleanliness when working with foods; using thermometers to monitor temperatures for time-temperature control for safety foods; and preparing, holding and serving food samples using best practices for keeping food safe.

References

1. Bullock S (2000) The economic benefits of farmers markets. Friends of the Earth [cited 2017 May15]. Available from: https://www.foe.co.uk/sites/default/files/downloads/farmers_markets.pdf
2. Andreatta S, Wickliffe W II (2002) Managing farmer and consumer expectations: a study of a North Carolina farmers market. Hum Organ 61:167–176
3. Smithers J, Lamarche J, Joseph AE (2008) Unpacking the terms of engagement with local food at the farmers market: insights from Ontario. J Rural Stud 24(3):337–350
4. Scallan E, Hoekstra RM, Angulo FJ, Tauxe RV, Widdowson M-A, Roy SL et al (2011) Foodborne illness acquired in the United States major pathogens. Emerg Infect Dis 17:715
5. U. S. Food and Drug Administration. FDA trend analysis report on the occurrence of foodborne illness risk factors in selected institutional foodservice, restaurant, and retail food

store facility types (1998–2008) [cited 2017 Jul 3]. Available from: https://wayback.archive-it.org/7993/20170113095247/http://www.fda.gov/downloads/Food/GuidanceRegulation/RetailFoodProtection/FoodborneIllnessRiskFactorReduction/UCM369245.pdf

6. Harrison JA, Gaskin JW, Harrison MA, Cannon JL, Boyer RR, Zehnder GW (2013) Survey of food safety practices on small to medium-sized farms and in farmers markets. J Food Prot 76(11):1989–1993

7. Simonne AH, Swisher ME, Saunders-Ferguson K (2006) Food safety practices of vendors at farmers markets in Florida. Food Prot Trends 26(6):386–392

8. Smathers SA, Phister T, Gunter C, Jaykus L, Oblinger J, Chapman B (2012) Evaluation, development and implementation of an education curriculum to enhance food safety practices at North Carolina farmers markets. Master's Thesis, North Carolina State University [cited 2017 Jul 3]. Available from: http://repository.lib.ncsu.edu/ir/bitstream/1840.16/8094/1/etd.pdf

9. Harrison JA (2014) Food safety and farmers markets [cited 2017 Jul 3]. Available from: http://www.foodsafetymagazine.com/magazine-archive1/junejuly-2014/food-safety-and-farmers-markets/

10. Park CE, Sanders GW (1992) Occurrence of thermotolerant campylobacters in fresh vegetables sold at farmers outdoor markets and supermarkets. Can J Microbiol 38:313–316

11. Scheinberg JA, Doores S, Cutter CN (2013) A microbiological comparison of poultry products obtained from farmers markets and supermarkets in Pennsylvania. J Food Saf 33:259–264

12. Scheinberg JA, Doores S, Cutter CN (2013) Food safety knowledge, behavior, and attitudes of vendors of poultry products sold at Pennsylvania farmers markets. J Ext 51(6):6FEA4

13. Worsfold D, Worsfold PM, Griffith CJ (2004) An assessment of food hygiene and safety at farmers markets. Int J Environ Health Res 14(2):109–119

14. Behnke C, Soobin S, Miller K (2012) Assessing food safety practices in farmers markets. Food Prot Trends 32(5):232–239

15. Choi J-K, Almanza B (2012) An assessment of food safety risk at fairs and festivals: a comparison of health inspection violations between fairs and festivals and restaurants. Event Manag 16(4):295–303

16. Terry L (2011) Oregon confirms deer droppings caused E. coli outbreak tied to strawberries. The Oregonian/OregonLive. August 17 [cited 2017 Jul 3] The Oregonian/OregonLive. Available from: http://www.oregonlive.com/washingtoncounty/index.ssf/2011/08/oregon_confirms_deer_droppings.html

17. Laidler MR, Tourdjman M, Buser GL, Hostetler T, Repp KK, Leman R, Samadpour M, Keene WE (2013) Escherichia coli O157:H7 infections associated with consumption of locally grown strawberries contaminated by deer. CID 57:1129–1134

18. Iowa Department of Public Health (2010) Iowa surveillance of notifiable and other diseases [cited 2017 Jul 3]. Available from: https://idph.iowa.gov/Portals/1/Files/CADE/IDPH_Annual_Rpt_2010_final.pdf

19. Ohlemeier D (2011) Iowa Salmonella outbreak traced to guacamole and salsa. The Packer [cited 2017 Jul 3]. Available from: http://www.thepacker.com/fruit-vegetable-news/fresh-produce-retail/iowa_salmonella_outbreak_traced_to_guacamole_and_salsa_122012119.html

20. Honish L, Predy G, Hislop N, Chui L, Kowalewska-Growska K, Trottier L, Kreplin C, Zazulak I (2005) An outbreak of E. coli O157:H7 hemorrhagic colitis associated with unpasteurized gouda cheese. Can J Public Health 96:182–184

21. Ellis A, Preston M, Borczyk A, Miller B, Stone P, Hatton B, Chagla A, Hockin J (1998) A community outbreak of Salmonella Berta associated with a soft cheese product. Epidemiol Infect 120(1):29–35

22. Green LR, Radke V, Mason R, Bushnell L, Reimann DW, Mack JC, Motsinger MD, Stigger T, Selman CA (2007) Factors related to food worker hand hygiene practices. J Food Prot 70(3):661–666

23. Pragle AS, Harding AK, Mack J (2007) Food workers perspectives on handwashing behaviors and barriers in the restaurant environment. J Environ Health 69:2733

24. Clayton D, Griffith C, Price P, Peters A (2002) Food handlers' beliefs and self-reported practices. Int J Environ Health Res 12(1):25–39
25. Green L, Selman C (2005) Factors impacting food workers' and managers safe food preparation practices: a qualitative study. Food Prot Trends 25:981–990
26. Hertzman J, Barrash D (2007) An assessment of food safety knowledge and practices of catering employees. Br Food J 109(7):562–576
27. Howes M, McEwen S, Griffiths M, Harris L (1996) Food handler certification by home study: measuring changes in knowledge and behavior. Dairy Food Environ Sanit 16:339–343
28. Howells A, Roberts K, Shanklin C, Pilling V, Brannon L, Barrett B (2008) Restaurant employees' perceptions of barriers to three food safety practices. J Am Diet Assoc 108:1345–1349
29. U. S. Food and Drug Administration. Current good manufacturing practice in manufacturing, packing, or holding human food. 21CFR.37 [cited 2016 Apr 1]. Available from: https://www.accessdata.fda.gov/scripts/cdrh/cfdocs/cfcfr/CFRSearch.cfm?fr=110.37
30. U. S. Department of Agriculture – Food Safety and Inspection Service (2016) Kitchen companion [cited 2016 Jul 3]. Available from: https://www.fsis.usda.gov/wps/portal/fsis/topics/food-safety-education/get-answers/food-safety-fact-sheets/safe-food-handling/kitchen-companion-your-safe-food-handbook/ct_index/!ut/p/a1/jZHRbsIgFIafxk-sKXZ2pu2uaLNrNdsZsYm8MbWkhtpwGcFv39EO90ugmXHHO9wfOB84xxblin7Jh-VoJi7eGcT7ZkSSb-NCZJNvWfyTz9WGYvcUzC1aMDNn8AaXBn_saKyH_55I4LHvQ-iXjQ475kVSKoaMG24RUyZL64NpjVAhQyruR1QzUqLjODcusahho5dwVTVStVgupO2FFyhErqeKWcJDbDX6BwtAHaYlnYrVcW_8Rrn588kvtvzNFiNZ0kakGx8CVzxeAJui3-ImmhaK46dtIlUEoRtZ85prrr29dmVhbW-eRmREDGe6FN7ABIDnJhmRayEBxjoJFyzuu3f68xrNiHzr1qGJfgGncMV4/

Chapter 10
Establishing a Food Safe Market: Considerations for Vendors at the Farmers Market

Renee R. Boyer and Lily L. Yang

Abstract Food safety at the market begins before the food reaches the market, continues while at the market, and extends after the market has ended for the day. The entire chain, from farmers to consumers, must be well-maintained in order to reduce the risk of contamination of a food product at the point of consumption. There are specific behaviors and practices that a market should maintain in order to establish a food safety culture and reduce the risk of foodborne illness. The five main contributing factors associated with foodborne illness are: (1) poor personal hygiene; (2) improper holding/time and temperature; (3) contaminated equipment/lack of protection of food from contamination; (4) inadequate cooking; and (5) food from unsafe sources. Controlling these factors can reduce the occurrence of foodborne illness. Attention to preventing these contributing factors is the cornerstone for establishing a market culture that values food safety. This chapter focuses on specific practices that a farmers market manager and vendors can follow to enhance safety.

Keywords Cross-contamination • Temperature control • Personal hygiene • Sanitation • Foodborne illness • Foodborne pathogens • Farmers market vendor

"Food safety culture" is an organizational culture of food safety that is made up of knowledge reflected in behaviors of the organization (in this case, the market) [1, 2]. For a farmers market manager to enhance the food safety culture within the market, recommended best practices must be practical, credible and cost effective [3]. The effectiveness can be influenced by the amount of support a farmer or producer receives on a particular guideline [3]. The recommendations listed in this chapter for what defines a food safe market must be bought into and operationalized by the

R.R. Boyer (✉) • L.L. Yang
Department of Food Science and Technology, Virginia Tech,
401-A HABB1 (0924), 1230 Washington Street SW, Blacksburg, VA 24061, USA
e-mail: rrboyer@vt.edu

© Springer International Publishing AG 2017   145
J.A. Harrison (ed.), *Food Safety for Farmers Markets: A Guide to Enhancing
Safety of Local Foods*, Food Microbiology and Food Safety,
DOI 10.1007/978-3-319-66689-1_10

market manager in cooperation with the vendors in order for them to be effective in reducing risk. There are several components that are essential for a food safe market mentality [3]. These include

- Transparency, or clearly communicated expectations
- Input from all parties (market and vendors)
- Recommendations for specific practices suggested to vendors that are:

  - science-based
  - flexible
  - continuously evolving
  - easy to understand.

Ultimately, the market should provide support to encourage agreement. This support can be infrastructure support, availability of a manager or trained personnel to answer questions, and/or training support.

To begin, the market should provide appropriate infrastructure for the vendors to be successful in keeping food safe (e.g., bathroom and handwashing facilities should be provided). The market should also have clear, written guidelines for the vendors that wish to sell at the market. Multiple parties (managers, vendors, educator etc.) should be involved in creation of these guidelines with reference to federal and state-mandated regulations. Having commitment from vendors to help create such guidelines will encourage buy-in from the vendors as a whole. The next step includes training of vendors in market practice recommendations to reduce food safety risk. The manager's role is to provide oversight to the recommended guidelines, but the most successful situation is one that involves vendors providing oversight collectively to each other.

There are five main practices/behaviors (in no order of significance) that are factors that contribute to causing foodborne illness cases and outbreaks [4]. These factors are:

1. Poor personal hygiene
2. Improper holding/time and temperature
3. Contaminated equipment/lack of protection of food from contamination
4. Inadequate cooking
5. Food from unsafe sources.

Of these, the most commonly identified risks found among vendors in farmers markets are improper holding/time and temperature, poor personal hygiene, contaminated equipment and lack of protection of products from contamination [5, 6]. The recommendations for establishing a food safe market focus on controlling those hazards. Regardless of where (local or imported) or how (organic or conventional) foods are produced these risk factors need to be recognized by food growers and handlers, and steps need to be put in place to reduce the risk of contaminated food reaching the consumer. In this chapter, outcomes of observational studies in farmers markets that identify risk factors present will be discussed and primary considerations for establishing a food safe market will be broken into recommendations for how markets should address each of these contributing factors. For a more detailed description of observational studies and tools for helping vendors and managers assess food safety risks on farms and in markets, refer to Chap. 9.

Poor Personal Hygiene

Proper health and hygiene of vendors and employees at the market is paramount to ensuring the health and safety of other employees, the safety of food items and, ultimately, the safety of consumers. Key ways to encourage good health and hygiene at a market include: (1) providing clean restrooms for vendors and customers, (2) providing adequate handwashing facilities for vendors and customers, (3) creating a culture that values handwashing, (4) encouraging minimal contact of foods with bare hands, and (5) encouraging the exclusion of sick employees and vendors from the market.

Providing Clean Restroom Facilities for Vendors and Customers

Providing adequate restroom facilities can be a challenge for farmers markets. Often markets are in a field or on a municipality's green space where there are no permanent structures in place. Observational data collected in North Carolina report 54% of markets provided restroom facilities at or near the market [6]; of those, 73% were considered clean [6]. Providing clean restroom facilities should be a priority for establishing a food safe market in order to ensure cleanliness and sanitation of vendors and shoppers at farmers markets. The restroom facilities should have working toilets, toilet paper, be located in close proximity to a well-stocked handwashing facility and be maintained for cleanliness; all while being kept separate from other areas.

Providing Adequate Handwashing Facilities for Vendors and Customers

Handwashing is paramount to food safety to prevent the spread of contamination and disease. Vendors should wash their hands frequently throughout the day when handling food. Observational data collected in North Carolina and Virginia report 27% and 40% respectively, of markets did not have handwashing stations or facilities throughout the market [6, 7]. Of the markets that did provide handwashing facilities, 87% did not post signs illustrating proper handwashing technique, and 22% did not provide soap [6]. A best practice for establishing a food safe market would be for all vendors to include handwashing facilities in their booth, and for the market to provide handwashing facilities for patrons; however, this may be difficult to achieve. For vendors specifically providing samples, preparing foods on site, or holding a food demonstration, handwashing stations are sometimes required by state law [8–10]. At a minimum, the market manager and vendor should be aware of and follow the local regulations requiring handwashing stations at sites where food preparation is occurring.

Handwashing stations can be easily built for any situation (even if bathrooms are not nearby). Primary components include a large container (5 gallon or 20 L) of water that has a spigot, a bucket to catch the water, hand soap in a pump dispenser, single-use paper towels and a trash receptacle. Figure 10.1 provides a diagram for what is required for an appropriate handwashing station. These are easy and inexpensive for vendors to set up at their booth at the market.

Creating a Culture That Values Handwashing

Handwashing can be done with cold, warm, or hot water. Hot water is effective in removing more oils; however, much of the effectiveness of handwashing is through the scrubbing action which dislodges dirt and microorganisms from the skin [11].

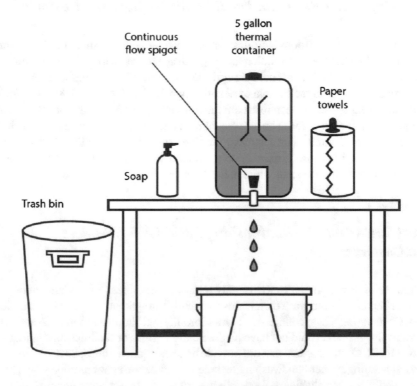

Fig. 10.1 Schematic diagram of components and set up of a temporary handwashing station. Courtesy Junyi Wu (junyiwu.com)

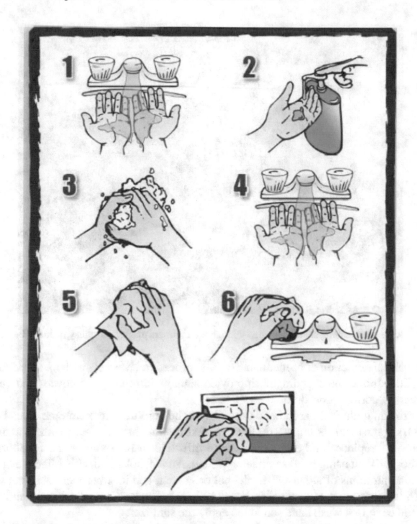

Fig. 10.2 Graphic depiction of proper handwashing procedure. Reprinted from Enhancing the Safety of Locally Grown Produce—Worker Hygiene. Boyer RR, Harrison JA, Gaskin JW, Harrison MA, Cannon J, Zehnder GW. 2012. University of Georgia Cooperative Extension Publication #FDNS-E-168-6

Recommendations for handwashing differ. Generally, all recommendations include (1) wetting hands, (2) applying enough soap to achieve a good lather, and (3) scrubbing hands vigorously, (4) rinsing, and drying. The time for scrubbing varies between 10 and 15 s [12–14] up to at least 20 s [15–17] depending on the source. The use of enough soap, mechanical action of scrubbing hands and drying with a clean single-use paper towel are the key actions contributing to removal of microorganisms [11, 17]. Single-use towels should be provided to dry hands as opposed to a hot air-dryer which can be contaminated with pathogens and deposit pathogens back onto the hands during the drying process [11, 17]. This is not an issue in most markets since they lack the infrastructure to provide hand dryers. Figure 10.2 shows

Fig. 10.3 Handwashing stations located throughout a farmers market. Courtesy of Judy Harrison

a graphical depiction of proper handwashing procedure. Markets with handwashing facilities should be encouraged to provide signage informing employees and customers of correct procedures.

Providing alcohol-based hand sanitizer can be helpful in reducing the spread of microorganisms. It is important, though, to stress that handwashing with soap and water is irreplaceable because it is more effective in removing some foodborne pathogens than using hand sanitizers [18, 19]. An alcohol sanitizer is effective only at concentrations of at least 60% alcohol or greater, and its effectiveness decreases with increased oils or dirt on the hands [20, 21]. The best use of a hand sanitizer would be to first wash hands and then apply the sanitizer.

A good strategy to encourage handwashing, would be for a market to provide and locate handwashing stations throughout the market (Fig. 10.3). This encourages handwashing for both vendors and shoppers by "nudging" [22]. Nudging is the act of coaxing, or gently encouraging people to make certain decisions. If the market had no handwashing stations, then no one would be able to wash their hands. By providing these facilities, the market may make a silent suggestion to individuals to use them. To be most effective, these handwashing stations should be well equipped and include signage educating the patrons on the proper way to wash hands.

The incorporation of handwashing stations at the market is only effective, if food handlers take the time to wash their hands when needed. The need for handwashing is often quite frequent, depending on the activity of the food handler. Figure 10.4 provides a diagram of when food handlers should wash their hands.

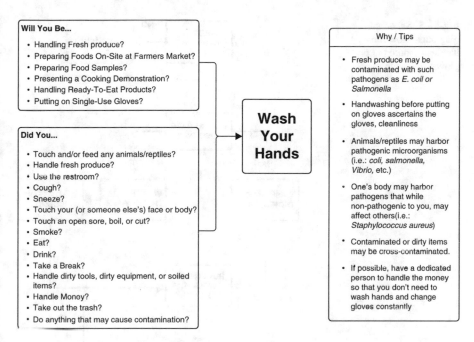

Will You Be...

- Handling Fresh produce?
- Preparing Foods On-Site at Farmers Market?
- Preparing Food Samples?
- Presenting a Cooking Demonstration?
- Handling Ready-To-Eat Products?
- Putting on Single-Use Gloves?

Wash Your Hands

Did You...

- Touch and/or feed any animals/reptiles?
- Handle fresh produce?
- Use the restroom?
- Cough?
- Sneeze?
- Touch your (or someone else's) face or body?
- Touch an open sore, boil, or cut?
- Smoke?
- Eat?
- Drink?
- Take a Break?
- Handle dirty tools, dirty equipment, or soiled items?
- Handle Money?
- Take out the trash?
- Do anything that may cause contamination?

Why / Tips

- Fresh produce may be contaminated with such pathogens as *E. coil or Salmonella*
- Handwashing before putting on gloves ascertains the gloves, cleanliness
- Animals/reptiles may harbor pathogenic microorganisms (i.e.: *coli, salmonella, Vibrio*, etc.)
- One's body may harbor pathogens that while non-pathogenic to you, may affect others(i.e.: *Staphylococcus aureus*)
- Contaminated or dirty items may be cross-contaminated.
- If possible, have a dedicated person to handle the money so that you don't need to wash hands and change gloves constantly

Fig. 10.4 Schematic diagram of when food handlers should wash their hands

Minimizing Bare Hand Contact

In addition to washing one's hands frequently, minimizing bare hand contact with food can also reduce the risk of foodborne illness. This can be done by using (a) gloves; (b) tongs; (c) spoons; and/or (d) hand papers/deli tissues when serving food. This is especially important when serving samples or allowing customers to serve out their own portions. Gloves provide a barrier between a food handler's hands and the food product. However, this can provide a false sense of security for the food handler and the customer if the gloves are misused. The same risks apply to using gloves to touch a food that apply to touching a food with bare hands. Gloves used for foodservice should be clean and replaced regularly. Without ensuring that they stay clean, the gloves can contribute to cross-contamination.

Single-use gloves are recommended for foodservice operations for sanitary purposes. Before putting on gloves, hands must be washed thoroughly for 10–20 s. If one has any open wounds, the affected area should first be covered with a bandage or barrier, then a single-use glove is pulled on atop that. Single-use gloves are designed for a single task only; upon completion of the task, both gloves must be discarded. During glove change, hands should be washed thoroughly, and a new pair of gloves should be used. Figure 10.5 provides a flow chart to indicate whether gloves are needed and the proper procedure for using them.

Fig. 10.5 Schematic
diagram of how food
handlers should put on
gloves and when to change
them

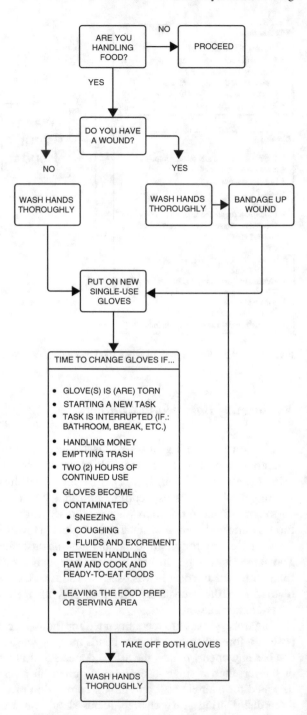

Exclusion of Ill Food Handlers

Sick vendors or employees should be excluded from handling food or any items that will come in contact with the food. If a worker exhibits any of the following symptoms, they should be excluded to minimize the risks of transmitting illness.

* Diarrhea, fever, vomiting, jaundice, or sore throat with fever
* Weeping or pus-filled cuts or wounds on any exposed/uncovered body parts (e.g., hands, arms, wrists, feet).

They also should be excluded if foodborne illness has been diagnosed in anyone they have been in contact with recently and/or with whom they live.

Improper Holding/Time and Temperature

In order to prevent foodborne illness, certain foods need to be held at temperatures that do not encourage microbial growth. Foods that require time and temperature control to ensure their safety are called TCS foods (Table 10.1). These include animal foods that are raw or heat-treated, plant foods that are heat-treated, raw seed sprouts, cut tomatoes, cut leafy greens, cut melons, and garlic in oil mixtures [12]. To maintain optimum safety, these foods either need to be held at or below 41 °F (5 °C) or at or above 135 °F (57.2 °C).

Temperature control is often seen at farmers markets, typically through the use of coolers filled with ice or chafing dishes to hold foods hot. Lack of electricity at most markets prevents the use of refrigeration. During observational data collection in markets, over 68% of vendors observed were not keeping a TCS food in a temperature controlled environment [7]. This alone is not terribly alarming since time is also required for microbial proliferation. Although the FDA Food Code states that TCS foods can be held without temperature control for up to 4 h or up to 6 h for refrigerated foods that are at 41 °F (5 °C) or colder when removed from refrigeration and the temperature does not exceed 70 °F (21.1 °C) within the 6 h period or hot foods are at least 135 °F (57.2 °C) when removed from temperature control, a more stringent two hour rule for farmers markets is a safer option. The USDA Pathogen Modeling Program for estimation

Table 10.1 Examples of time-temperature control for safety (TCS) foods observed in farmers markets

Eggs	Raw chicken	Raw beef
Milk	Yogurt	Cut cantaloupe
Cut watermelon	Cut spinach	Cut lettuce mixes
Cooked meat (e.g., barbeque, hamburgers)	Cooked vegetables	Sliced tomatoes
Diced tomatoes	Cooked casseroles	Cooked soups

of growth uses an ambient temperature of 65 °F (18.3 °C). Ambient temperatures at outdoor farmers markets may well exceed 65 °F (18.3 °C) in summer months allowing food temperatures to change rapidly. Consumer recommendations from USDA, FDA and all nationwide food safety education campaigns specify that foods should not be held in the temperature danger zone (between 40 °F and 140 °F; 4.4 °C and 60 °C) for more than two hours or one hour if the ambient temperature is above 90 °F (32.2 °C). Therefore, the safest option for using time alone as a food safety control at a farmers market would be to use the consumer two hour rule (or one hour at surrounding temperatures above 90 °F or 32.2 °C) rather than rules for restaurant and retail foodservice environments from FDA [12]. However, the best practice for a food safe market is to hold foods at the recommended holding temperatures. Hold hot foods at or above 135 °F (57.2 °C), and hold cold foods at or below 41 °F (5° C) (Fig. 10.6) [12].

Inadequate cooling is less of an issue in a farmers market because most foods that are served there are going to be held hot, or cold, but not cooked, cooled, and stored at the market for use at a later time. However, correct cooling of a product before reaching the market is an important food handling practice since temperature abuse can allow microorganisms in a food to multiply to high levels. It is very important to rapidly cool foods for storage to prevent any microbial growth. This could be important for a market vendor if they are preparing a TCS food in their home kitchen to take and heat up later to serve at the market. One example might be a soup that might be made ahead of time and then warmed up. To maintain the safest cooling, foods should be cooled from 135 °F (57.2 °C) to 70 °F (21.1 °C) within two hours, reaching a final cold holding temperature of 41 °F (5 °C) within a total of 6 h [12]. There are a variety of methods to ensure that a food is cooled quickly under these parameters. Making sure that the maximum surface area of the product has exposure to cold temperatures is the primary way. This can be done by dividing the product into smaller portions, or placing it in shallow pans before placing it in the refrigerator. Additional methods include adding ice as an ingredient, using rapid cooling equipment, or using an ice wand to stir the product, reducing the temperature [12]. While this cooling process is unlikely to occur at the market, market managers should educate vendors on correct cooling practices through training materials.

Temperature control of TCS foods does not start at the point of sale. Vendors need to be trained on ways to maintain correct temperature storage during transportation as well. Foods that need to be refrigerated should be transported in a cooler filled with potable ice and monitored to ensure that the product stays below 41 °F (5 °C). Foods that will be served or prepared at the market should be transported under refrigeration before cooking at the market.

Creating a market culture of thermometer use can significantly reduce risk. All vendors should be encouraged to have a properly calibrated food thermometer on site and use it to ensure that foods stay within appropriate temperatures while at the market. This is especially important during summer months where the ambient temperatures of some markets can reach above 100 °F (37.8 °C)

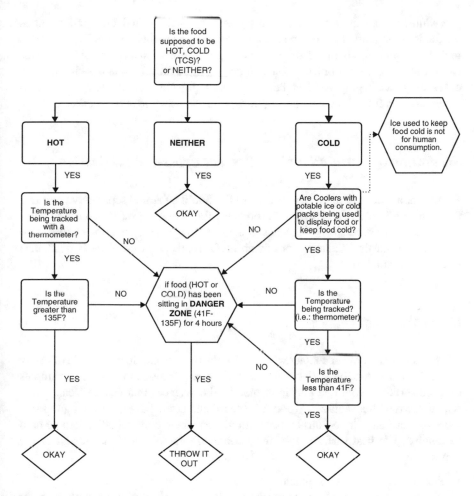

Fig. 10.6 Schematic diagram of how to hold hot/cold foods at the market and when to discard

influencing cold holding temperatures. The thermometer should be one that can be easily calibrated to ensure accuracy. Thermometers should be cleaned and sanitized between uses (even in the same use) to avoid potential cross-contamination.

Contaminated Equipment/Lack of Protection of Foods from Contamination

The third of the factors that contribute to foodborne illness outbreaks which may pertain to farmers markets is contaminated equipment/lack of protection of foods from contamination. Preventing cross-contamination and practicing good sanitizing procedures is essential for safely serving food. This includes reducing opportunities for foods at the market to become contaminated from another source. Key practices include keeping foods off of the ground to prevent dirt, dust or splashing liquids from contacting them. For foods that can be packaged for sale (like breads, cookies, cakes etc.), packaging is ideal because it can prevent any unintentional cross-contamination from occurring during transportation or while at the market.

Separation of Certain Foods to Prevent Cross-Contamination

Foods that are considered ready-to-eat (RTE) should be stored separately from raw foods that will require cooking. RTE foods are foods that are safe to consume without any further processing [12]. A best practice is to use different coolers, refrigeration units or plastic containers for raw and RTE foods to ensure there is no chance of cross-contamination.

Cleanable Surfaces

All surfaces that come in contact with food at the market should be capable of easily being cleaned. Cleanable surfaces are generally non-porous solid surfaces. Examples of good surfaces would be plastic or metal solid surfaces that can be cleaned and wiped down with a sanitizer. Wood, cardboard and straw baskets are not good surfaces because they are porous and contain niches where contamination can be hard to remove. The best surface types for a food safe market would be for all of the vendors to use:

- Plastic tables that are cleanable
- If using a wooden table, making sure that a clean table cloth is used on the table
- Plastic, cleanable bins to hold loose food items like produce
- If using cardboard boxes, use a plastic or paper liner that can be disposed of following the market
- If using baskets to display product, line with a clean dish towel or paper towels.

To increase sustainability, vendors often like to use reuse bags for the products they sell. Previously used grocery bags or other plastic bags that once held food are not recommended because they may contaminate the new uncontaminated product with the remnants of their previous contents. Garbage bags are also not recommended to be used to contain or transport foods because they are not food-grade. Some garbage bags are lined with mold inhibitors and other chemicals that should not come in contact with food. The best practice would be for vendors to used new clean, single use bags. It is also ideal for vendors to separate product types. For example, place meat and ready-to-eat vegetables in separate bags to prevent cross-contamination.

Cloth bags are used by many patrons, but they should be encouraged to wash them on a regular schedule. Research has shown that most cloth grocery bag users do not clean their bags [23]. When tested, no foodborne pathogens were found present, but coliforms were found in half of reusable grocery bags, and *Escherichia coli* was found in 8% [23]. This does not necessarily translate to a food safety risk, but it demonstrates that these bags should be cleaned on a regular schedule. Most can be washed in a washing machine or hand-washed with warm water and detergent (depending on the bag). In addition to providing separate bags and ice to keep vegetables, meats, poultry, seafood, and RTE foods separated and cool, vendors and farmers market managers should remind patrons to wash their cloth bags to prevent cross-contamination of goods.

Use of Sanitizers

Surfaces should be sanitized when setting up the booth for the market, and when taking the booth/display down at the end of the day. A simple bleach solution is the most cost-effective sanitizer that can be used for this purpose. The sanitizing agent in bleach is sodium hypochlorite, more commonly known as chlorine. Chlorine chemistry is complicated and its effectiveness depends on many factors such as pH and temperature of the water [24]. As the water becomes dirty with organic material, the sanitizing effectiveness also decreases [24].

A solution with no more than 200 parts per million is allowed for food contact surfaces [12]. To make a 200 ppm bleach solution, mix one tablespoon (14.8 mL) of plain, unscented household bleach (5.25–6% hypochlorite) with one gallon (3.8 L) of water. This solution can be added to a spray bottle for ease of use. To sanitize market surfaces, the vendor should spray (or wipe) solution on and allow it to air dry for maximum sanitation. If concentrated bleach (8.25% hypochlorite) is used, then use two teaspoons (9.9 mL) per one gallon (3.8 L) of water to make a 200 ppm solution. For maximum effectiveness, the solution should be tested with commercially available chlorine test strips to ensure the proper concentration. These are relatively inexpensive and can be purchased at a variety of different retail outlets including restaurant supply stores.

Depending on the state or location, some vendors at a market may be required to have a three compartment sink for cleaning and sanitizing utensils and other food contact items during service. This is most commonly required for sale of foods that are prepared on site at the market. To create a simple three compartment sink for use at the market, a vendor simply needs three plastic tubs. One labeled wash, one labeled rinse and one labeled sanitize. Figure 10.7 shows this type of set up.

The same concentration of bleach solution can be used for the third "sanitize" container. Generally it is recommended for utensils or other food contact items to be

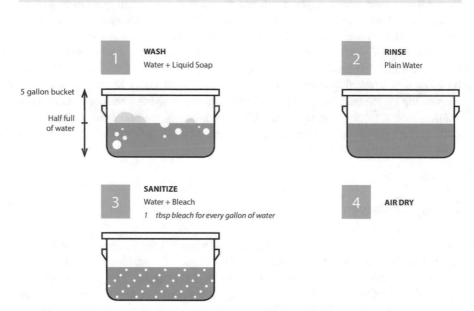

EXAMPLE OF UTENSIL WASHING SET-UP

Fig. 10.7 Schematic diagram of a three compartment sink set up at an outdoor market. Courtesy Junyi Wu (junyiwu.com)

submerged in the sanitizer for at least 1 min to be effective. Figure 10.8 shows a flow diagram for how to properly use the three compartment sink for cleaning and sanitizing. This can be set up on a table behind the food preparation area for ease of access.

Exclusion of Animals

Due to the community focus and open nature of farmers markets, patrons enjoy bringing their pets to the market. Ideally, a food safe market would exclude pets from the areas where food is being sold. Animals, even pets, can carry pathogens like *E. coli* and *Salmonella* and even parasites. Exclusion of animals can be a controversial topic. Patrons tend to be very vocal about wanting to bring their dogs with them to the market. In most locations, the decision to exclude animals is a decision specific to each market. To be effective, this decision should be driven by the culture of the market and its vendors. However, rather than having a policy to exclude animals, some market managers choose to use signage to encourage patrons to leave their pets at home. In other locations, animals are either prohibited from entering the market or must be kept on leashes and a certain distance away from food displays (Fig. 10.9).

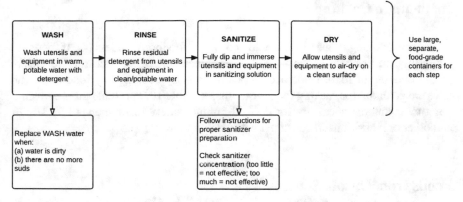

Fig. 10.8 A schematic diagram outlining the wash, rinse, sanitize procedure for using a three-compartment sink at a farmers market

Fig. 10.9 Sign explaining regulations about pets in the market

Inadequate Cooking

Inadequate cooking is one of the leading causes of foodborne illness [4]. If foods are being prepared on-site at the market, thermometer usage is necessary to ensure proper cooking of food. Undercooking or improperly cooking any food items can have the potential for causing foodborne illness. Table 10.2 presents minimum end-point temperatures necessary for cooked product safety recommended by USDA and other organizations in consumer education initiatives.

Foods from Unsafe Sources

To create a food safe market, a market manager needs to be sure that the foods that are arriving at the market for sale by the vendors are as safe as possible upon arrival. This can be difficult since the manager has little control over the conditions that the food has been grown in or processed in prior to arrival. The most important thing is to be sure that the foods being sold are legal by federal and state guidelines. Different states have different laws governing which foods fall under inspection from the state. Table 10.3 provides a brief summary of some of the inspection requirements for specific commodities. These requirements are state specific in many cases, therefore a best practice for vendors or market managers is to understand the specific requirements for their state.

Not only is it important for market vendors to follow the regulations related to the specific produce they are selling, but also vendors should be growing, harvesting, packing and processing their products in a safe manner. Often small direct market local growers and processors lack access to education and training related to best practices. To ensure that all vendors have appropriate training, a market manager could train vendors themselves, or require that vendors to attend a third-party training where they obtain a completion certificate which can be used to confirm that they have completed appropriate training. State Cooperative Extension Services can often provide this type of training for small direct market vendors.

Table 10.2 Recommended endpoint cooking temperatures to ensure safety of foods

Food	Temperature (°F/°C)
Poultry	165/73.9
Ground meat (beef, pork, veal, lamb)	160/71.1
Egg dishes	160/71.1
Fish, steaks, roasts (beef, pork, veal, lamb)	145/62.8 (with 3 min rest time)

Adapted from USDA Food Safety and Inspection Service. Available from: https://www.fsis.usda.gov/wps/portal/fsis/topics/food-safety-education/get-answers/food-safety-fact-sheets/safe-food-handling/safe-minimum-internal-temperature-chart/ct_index

Table 10.3 Brief summary of some of the inspection requirements for specific commodities at a farmers market

Food product	Stipulations for sale at market
Meat (cattle, hogs, sheep, goats, chickens, turkeys, ducks, geese, guineas, ratites and squabs)	• USDA inspection at the time of slaughter is required for sale (see exceptions for poultry) • Specific federal mandated labeling is required
Poultry (exceptions)	• 1000 bird exemption—vendor may sell poultry within the state that they raised, slaughtered and processed on their own property. Must have proper labeling • 20,000 bird exemption—vendor may sell poultry under producer/grower exemption, poultry that they slaughter and process on their own property
Dairy products	• All dairy must be inspected • Sale of raw (unpasteurized) milk depends on state
Eggs	• No inspection needed if vendor selling less than 150 dozen of own eggs/week • No inspection needed if vendor selling less than 60 dozen of another producer's eggs/week
Fresh produce	Falls under Produce Safety Rule [25]. Growers exempt if: • Vendor makes less than $25,000 in produce sales annually (averaged over 3 years) • Vendor selling exempt produce commodity • If qualified exemption criteria are met for those selling more than $25,000. Vendors combined food sales are less than $500,000 annually (averaged over 3 years) and >50% of sales are in the direct market (e.g., Farmers market, restaurants or retail within the same state and not more than 275 miles away)
Acidified foods	Depend on state
Low acid foods	Fall under federal regulation and are not allowed to be sold in direct market venues which include farmers market

Special Considerations Specific to Farmers Markets

Providing Samples to the Public

Farmers market vendors often like to offer samples to customers to promote their products. This contributes to some of the charm of shopping at a farmers market. However, care needs to be taken to ensure that samples are prepared and offered in a safe manner. Ideally, samples should be single-serving sized only and prepared prior to the market. This way there is little to no food preparation occurring on site. To create single servings, two ounce (60 mL) single serve

Table 10.4 Recommended food safety practices for offering samples at farmers markets

Recommended best practices for offering samples
Make samples single serve
• Use toothpicks or single serve disposable utensils
• Prepackage items, for example: 2 oz cups (60 mL) with lids prepared at home
Hold samples requiring refrigeration on ice to keep cold during sampling
Hold cooked samples above 135 °F (57.2 °C)
Have a lid or cover over the samples to prevent insects
Have a dedicated server to offer the samples instead of customers serving themselves
Have a handwashing station on site
Have a three compartment sink on site to clean and sanitize any utensils being used
Offer samples on a clean table and store samples at least 6 in. (15.2 cm) off the ground

cups with lids are a good strategy. Samples should be covered and protected against insects, dust, and other potential physical hazards as well as microbial contamination. While patrons will likely want to choose their own sample, it is best not to allow patrons to take their own samples. Serve individual samples to patrons as desired. This reduces the possibilities for cross-contamination as the situation is controlled by the vendor. Table 10.4 lists some of the best practices for a vendor to follow when providing samples to the public.

While preparing single serve samples prior to arriving at the market is considered a best practice, some vendors prefer preparing samples while at the market. This can be done, but requires a few more steps. Providing samples that are prepared on site must follow the same considerations as any food preparation that occurs at the market. Many states require a handwashing station and three compartment sink set up for adequate hygiene, cleaning and santiation during food preparation and serving. Figure 10.10 provides a flow chart of steps for vendors to follow when preparing samples at the market for the public.

Summary

This chapter has provided recommendations for best practices when establishing a food safe market. The most important factor is the creation of a supportive culture that provides food safe infrastructure as well as education and training for vendors. Table 10.5 provides a summary checklist of practices that contribute to establishing a food safe market. This checklist can be used by market managers or vendors to

Fig. 10.10 Flowchart of best practice steps to follow when serving samples at a farmers market

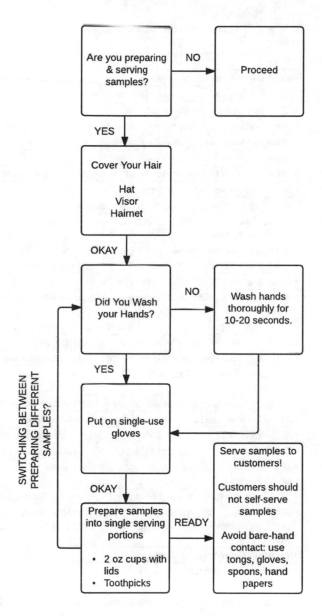

Table 10.5 Summary checklist of practices that contribute to establishing a food safe farmers market

Practice	Yes	No
Preventing poor personal hygiene		
• Handwashing stations are provided and adequately equipped		
• Signs displaying proper way to wash hands		
• Hand sanitizer for additional hygiene if provided (not to replace handwashing)		
• Bathroom facilities are provided and adequately equipped		
• Market has a sick worker policy		
• Market vendors on the importance of cleanliness and handwashing at the market		
• Bare hand contact is discouraged		
Preventing improper holding time/temperature		
• Market vendors are trained on TCS foods, and how to properly hold foods at certain temperatures		
• Refrigeration units or coolers with potable ice to hold TCS foods at the market		
• Hot foods are held above 135 °F (57.2 °C) while at the market		
• Coolers with potable ice are used to hold cold foods during transportation		
• All vendors use calibrated thermometers to monitor temperatures of hot and cold foods		
• Records are kept to ensure holding foods at correct temperatures		
• Foods held at ambient temperature for 4–6 h are thrown out		
Preventing contaminated equipment and cross-contamination		
• Market vendors are trained on and adhere to proper cleaning and sanitation protocol		
• Market vendors are trained on and adhere to methods to prevent cross contamination		
• Animals are excluded from areas of the market selling food		
• All surfaces used to display foods, or touching foods are cleanable		
• Cardboard boxes, wooden crates and other porous surfaces are not use		
• Market surfaces (include vendor stalls) are cleaned and sanitized at the beginning of the market and at the close of the market		
• RTE foods are kept separated from raw foods that require cooking to be safe		
• Vendors selling prepared foods or providing samples have a handwashing station and three compartment sink to clean/sanitize utensils and other food preparation items		
Preventing inadequate cooling		
• Market vendors are trained on, and adhere to safe cooling practices to use		
Preventing the sale of foods from unsafe sources		
• Market vendors are trained on and adhere to appropriate regulations for the products they sell		
• All produce sold adhere to state and federal regulations		
• Vendors are evaluated to ensure that they are using safe practices on their farm/ processing facility before becoming approved to sell at the market		

determine whether or not they are following recommended best practices to enhance product safety. Additional information and tools for assessing risky practices in farmers markets are available in Chap. 9.

References

1. Yiannas F (2009) Food safety culture: creating a behavior-based food safety management system. Springer Science, New York
2. Powell DA, Jacobs CJ, Chapman BJ (2011) Enhancing food safety culture to reduce rates of foodborne illness. Food Control 22(6):817–822
3. Chapman B, Powell D (2005) Implementing on-farm food safety programs in fruit and vegetable cultivation [cited 2016 May 10] Available from: http://citeseerx.ist.psu.edu/viewdoc/download?doi=10.1.1.381.8509&rep=rep1&type=pdf
4. U.S. Food and Drug Administration (2000) Report of the FDA retail food program database of foodborne illness risk factors [cited 2016 May 10]. Available from: http://www.fda.gov/downloads/Food/GuidanceRegulation/UCM123546.pdf
5. Harrison JA, Gaskin JW, Harrison MA, Cannon JL, Boyer RR, Zehnder GW (2013) Survey of food safety practices on small to medium-sized farms and in farmers markets. J Food Prot 76(11):1989–1993
6. Smathers SA, Phister T, Gunter C, Jaykus L, Oblinger J, Chapman B (2012) Evaluation, development and implementation of an education curriculum to enhance food safety practices at North Carolina farmers' markets. Master's Thesis, North Carolina State University [cited 2016 March 16]. Available from: http://repository.lib.ncsu.edu/ir/bitstream/1840.16/8094/1/etd.pdf
7. Pollard S, Boyer R, Chapman B, Ponder M, Rideout S, Archibald T (2016) Identification of risky food safety practices at southwest Virginia farmers' markets. Food Prot Trends 36(3):168–175
8. Oregon Dept. of Agriculture (2013) Food safety at farmers' markets information & guidelines [cited 2017 Apr 20]. Available from: https://www.oregon.gov/ODA/shared/Documents/Publications/FoodSafety/FarmersMarketsFoodSafety.pdf
9. Hofman C, Dennis J, Gilliam AS, Vargas S (2007) Food safety regulations for farmers' markets. Purdue extension EC-740 [cited 2017 Apr 20]. Available from: https://www.extension.purdue.edu/extmedia/ec/ec-740.pdf
10. Nebraska Dept. of Agriculture (2016) Food safety and regulation requirements for farmers' markets and craft shows [cited 2017 Apr 20]. Available from: http://www.nda.nebraska.gov/publications/foods/food_safety_farmers_markets_craft_shows.pdf
11. Snyder P (1998) Hand washing for retail food operations – a review. Dairy Food Environ Sanit 18(3):149–162
12. U.S. Food and Drug Administration (2013) Food code [Internet]. College Park, MD [cited 2016 March 17]. Available from: http://www.fda.gov/downloads/Food/GuidanceRegulation/RetailFoodProtection/FoodCode/UCM374510.pdf
13. National Restaurant Association (2014) ServSafe® food handler guide, 6th edn. National Restaurant Association Solutions, Chicago, IL
14. National Restaurant Association (2014) ServSafe® manager, 6th edn. National Restaurant Association Solutions, Chicago, IL
15. Partnership for Food Safety Education [Internet]. Washington DC. Clean – wash hands and surfaces often [cited 2016 Apr 27]. Available from: http://www.fightbac.org/clean-1/
16. Centers for Disease Control and Prevention (2016) [Internet] Atlanta, GA. Wash your hands [cited 2016 Apr, 16]. Available from: http://www.cdc.gov/features/handwashing/

166   R.R. Boyer and L.L. Yang

17. Montville R, Chen Y, Schaffner DW (2002) Risk assessment of hand washing efficacy using literature and experimental data. Int J Food Microb 73(2–3):305–313
18. Blaney DD, Daly ER, Kirkland KB, Tongren JE, Kelso PT, Talbot EA (2011) Use of alcohol-based hand sanitizers as a risk factor for norovirus outbreaks in long-term care facilities in northern New England: December 2006 to March 2007. Am J Infect Control 39(4):296–301
19. Oughton MT, Loo VG, Dendukuri N, Fenn S, Libman MD (2009) Hand hygiene with soap and water is superior to alcohol rub and antiseptic wipes for removal of *Clostridium difficile*. Infect Control Hosp Epidemiol 30(10):939–944
20. Kampf G, Kramer A (2004) Epidemiologic background of hand hygiene and evaluation of the most important agents for scrubs and rubs. Clin Microbiol Rev 17(4):863–893
21. Todd EC, Michaels BS, Holah J, Smith D, Greig JD, Bartleson CA (2010) Outbreaks where food workers have been implicated in the spread of foodborne disease. Part 10. Alcohol-based antiseptics for hand disinfection and a comparison of their effectiveness with soaps. J Food Prot 73(11):2128–2140
22. Oldfield K (2016) The social aspects of hand washing in American restaurants: an administrative approach to reducing public and private health care costs. Adm Soc 49(5):753–771. doi:10.1177/0095399716638121
23. Williams DL, Gerba CP, Maxwell S, Sinclair RG (2011) Assessment of the potential for cross-contamination of food products by reusable shopping bags. Food Prot Trends 31(8):508–513
24. White GC (1986) Handbook of chlorination, 2nd edn. Van Nostrand Reinhold, New York, NY, pp 150–213
25. U. S. Department of Health and Human Services (2015) Federal register. 21 CFR Parts 11, 16 and 112. Standards for growing, harvesting, packing and holding of produce for human consumption; final rule [Cited 2016 May 9]. Available from: https://www.gpo.gov/fdsys/pkg/FR-2015-11-27/pdf/2015-28159.pdf

Index

© Springer International Publishing AG 2017
J.A. Harrison (ed.), *Food Safety for Farmers Markets: A Guide to Enhancing
Safety of Local Foods*, Food Microbiology and Food Safety,
DOI 10.1007/978-3-319-66689-1